CONTEMPORARY CASE STUDIES

Cities & Urbanisation

Michael Witherick & Kim Adams

Series Editor: Sue Warn

Philip Allan Updates, an imprint of Hodder Education, part of Hachette UK, Market Place, Deddington, Oxfordshire OX15 0SE

Orders

Bookpoint Ltd, 130 Milton Park, Abingdon, Oxfordshire OX14 4SB
tel: 01235 827720
fax: 01235 400454
e-mail: uk.orders@bookpoint.co.uk

Lines are open 9.00 a.m.–5.00 p.m., Monday to Saturday, with a 24-hour message answering service. You can also order through the Philip Allan Updates website: www.philipallan.co.uk

© Philip Allan Updates 2006

ISBN 978-1-84489-220-4

Contents

Introduction

Part 1: A global view 1

Part 2: The urbanisation pathway: early phases 10

Part 3: The urbanisation pathway: later phases 23

Introduction

With more than half of the global population now residing in towns and cities, we live in an urban world. Two hundred years ago, only 1 in 50 people lived in towns and cities, and rural areas dominated. More recently, particularly since the 1960s, the world has been transformed by the process of urbanisation. As countries have progressed along what we might call the **urbanisation pathway** (Figure 1A), their populations have become increasingly concentrated in an extending network of towns and cities.

Urbanisation has significantly changed where people live. At the same time, a distinctive environment has emerged within those growing towns and cities. This urban environment has a profound impact on how its inhabitants live — what they do for a living, their homes, their lifestyles, their values and aspirations. Together, the urban environment and its inhabitants have a major impact on the wider natural environment. This is frequently referred to as their **ecological footprint**.

About this book

The underlying theme of this book is dynamism and change. It focuses on six aspects (Figure 1B):

- **Part 1** is a global stocktaking. The aim is to establish some of the broad dimensions and outcomes of urbanisation today.
- **Part 2** examines the urban situation in parts of the world less advanced along the development and urbanisation pathways compared with countries such as the UK (Figure 1A). It was the UK that spearheaded the rise of an urban world in its now largely westernised form. The focus here is on countries falling within the World Bank **development group** sequence, ranging from 'low income' to 'upper-middle income'.
- For those countries further along the development pathway — the 'high-income' countries — the process of urbanisation appears to take a U-turn. In its early stages, urbanisation involves mainly **centralisation**. However, **part 3** demonstrates that the basic character shifts towards **decentralisation**. Towns and cities loosen up their structures as the growth impetus shifts from the largest cities to middle- and lower-order urban settlements. Many believe this change is 'anti-urban', which is wrong — the shift simply represents the wider spread of urbanisation and its associated lifestyles. The use of the term 'counterurbanisation' is partly responsible for this common misunderstanding.
- In **part 4**, the spotlight remains on MEDCs. It assesses the various processes moulding the character of today's urban environment. While decentralisation

continues to add new growth to the margins of towns and cities, the existing built-up area is being changed by a variety of processes and management interventions (by planners, investors etc.) The urban environment is being transformed contin-uously. For example, efforts are being made to regenerate older parts of towns and cities (**reurbanisation**). The density of development in suburban areas is also being raised (**suburban intensification**). Experiments are being made with new forms of urban living, such as **edge cities** and **gated communities**. Government direction, intent on controlling urban form and change, is making its mark. This is helping to generate a dynamism within the built-up area or urban environment that is economic, social and cultural.

- **Part 5** investigates and illustrates some of the more important issues associated with today's urban environment. Most of these issues have a bearing on the quality of life, and include housing, transport, services and personal security. They challenge those who are responsible for managing the urban environment and ensuring a more sustainable urban world.
- The search for urban **sustainability** and more eco-friendly cities is the theme of **part 6**. These aspects are motivated by an increasing awareness of the ecological footprints of towns and cities. Urban decision makers need to consider the awesome technology that marks the latest wave of the Industrial Revolution. Will this new technology create a radically different type of built environment? Might it allow us to do away with cities altogether?

Thus, the six major themes of urbanisation geography are:
- the global situation
- the process of centralisation (agglomeration) in less developed countries
- the process of decentralisation in more developed countries
- the dynamic nature of the urban environment
- the issues created by that dynamism
- the urban future

Although these six components are treated separately in this book, they are closely interlinked (Figure 1B). This point is reinforced by the fact that a number of the case studies relate to nine recurring countries as well as to some of their cities: Bangladesh (Dhaka), Ethiopia (Addis Ababa), China (Shanghai), Mexico (Mexico City), Iraq (Baghdad), Russia (St Petersburg), Singapore (a single city state), the USA (Los Angeles and Chattanooga) and the UK (London, Milton Keynes, Southampton and Bradford). Each of these countries represents one of the main phases along the urbanisation pathway. It may seem odd that the USA is shown to be behind the UK on this pathway. However, although it enjoys a higher per capita **gross national income (GNI)**, it is significantly less urbanised.

Advice is given throughout on making the best use of case studies at A-level. Some of the case studies are followed by *Using case studies* boxes. Most of these show how a particular case study might be useful for answering a specific question; some invite you to attempt an exercise based on one or more of the case studies. You will also find some tips on tackling tasks that are commonly set in examinations, such as:
- describing the main features of maps showing the distribution of towns and cities at different spatial scales
- identifying the spatial character of a city's internal structure

- extracting the basic message conveyed by a statistical table
- producing and annotating a sketch map or diagram

The final part of the book consolidates much of this advice and gives some additional tips on making the most of your case study material in the examination.

Figure 1 **(A)** *The urbanisation pathway and development groupings*
(B) *The content of this book in relation to the urbanisation pathway*

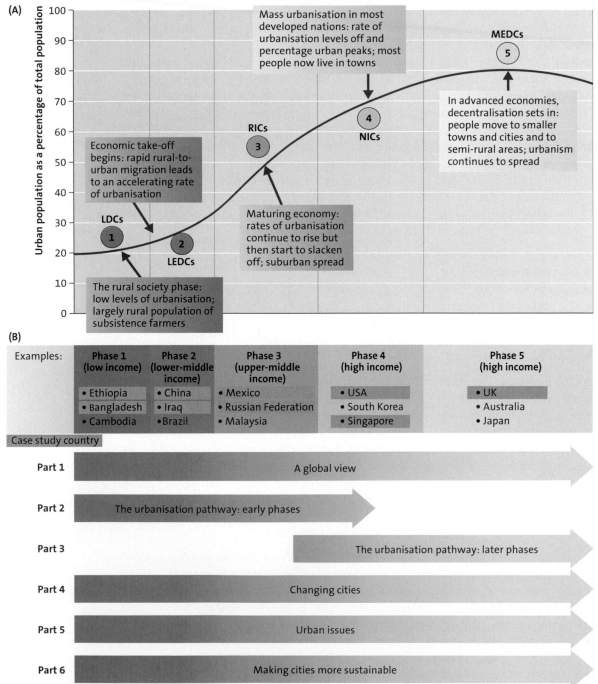

Key terms

Affordable homes: housing made accessible to poorer and new households. It mainly comprises social housing made available at relatively low rents, and low-cost housing for sale to owner-occupiers through low-interest mortgages.

Anchor development: a key activity, such as a department store, that is likely to attract other shops and therefore helps to guarantee the success of a new shopping mall.

Bottom up: where a local community initiates a particular development.

Brown Agenda: part of the international drive to make cities and urbanisation more sustainable. It is mainly concerned with the living environment and quality of life of slum dwellers.

Brownfield: land that has been previously used, abandoned and now awaits new use.

Central business district (CBD): the commercial centre of a town or city.

Centralisation: the tendency for people and economic activities to become concentrated at specific points.

Centrifugal: the outward movement of people and activities.

Centripetal: the inward movement of people and activities.

Cloning: a process whereby places lose their individuality and become more alike.

Comparative advantage: the ability to produce something at a lower cost than competitors, largely through specialisation.

Consolidation: the upgrading of shanty towns and squatter settlements and their official integration into the built-up area of a town or city.

Conurbation: a continuous urban area formed by the growing together of once-separate towns and cities.

Counterurbanisation: the movement of people and employment from major cities to smaller settlements and rural areas located just beyond the city, or to more distant, smaller cities and towns.

Decentralisation: the outward movement of people and economic activities from established centres.

Deindustrialisation: the absolute or relative decline in the importance of manufacturing in the economy particularly in MEDC cities.

Deprivation: when an individual's wellbeing falls below a level generally regarded as a reasonable minimum. Measuring deprivation usually relies on indicators relating to employment, housing, health and education.

Development: the increased use of resources and technology that leads to a rise in the standard of living of a country.

Development groups: World Bank (2005) classification of countries based on per capita gross national income (GNI). Four groups are recognised: low income (US$735 or less); lower-middle income (US$736–2935); upper-middle income (US$2936–9075); and high income (US$9076 or more).

Downtown: a North American term for the central business district.

Ecological footprint: the impact people have on the environment, principally through the consumption of resources and the disposal of waste.

Economic globalisation: the process whereby the economies of the world are moving closer together and becoming integrated.

Edge city: the result of urban processes, leading to parts of the suburbs becoming more city-like through the agglomeration of offices, factories and large shopping complexes.

Exclusion: the shutting out of particular groups of people (e.g. the poor, ethnic minorities) from full participation in society or from particular areas.

Filtering: the process by which ageing housing becomes occupied by lower and poorer groups in the social hierarchy as the upwardly mobile move out.

Flagship development: a large prestigious scheme, often designed to kick-start the development or improvement of a larger area.

Gated community: an area of wealthy private housing with a secure perimeter wall or fence with controlled entrances for residents, visitors and their cars.

Gentrification: the movement of middle-class people back into run-down, inner-city areas, resulting in an improvement of the housing stock and image.

Ghetto: a residential area that is largely occupied by one ethnic or cultural minority group.

Global city: one of the world's leading cities — a major node in the complex economic networks being produced by globalisation. The influence of global cities (e.g. London, New York, Tokyo) is linked to their provision of financial and producer services.

Greenfield: land that has not been subject to any urban development.

Gross domestic product (GDP): the total value of goods and services produced by a country in 1 year but excluding net income from abroad.

Gross national income (GNI): previously known as gross national product — the total value of goods and services produced by a country in 1 year, including net income (interest, dividends etc.) from other countries. Usually expressed in per capita terms.

Heritage: any feature (natural or manmade) inherited from the past.

Image: the perceived reputation and appearance of a town or city.

Informal economy: made up of employments that are not officially recognised. They are commonly found in LEDCs where people have to find work for themselves, often illegally, on the streets and in small workshops.

Infrastructure: the amenities and services basic to daily life. They include water supply, sewage disposal, schools, healthcare and roads.

Inward investment: injections of capital made by foreign companies in order to extend their business interests and profits.

Least developed country (LDC): one of the poorest countries in the world — where per capita GNI is less than US $750 per annum.

Less economically developed country (LEDC): a relatively poor country in the early stages of development. The term is currently applied to most countries in Africa, Asia (excluding Japan), Latin America, the Caribbean, and the regions of Melanesia, Micronesia and Polynesia.

Mega-city: a city or urban agglomeration with a population of 10 million or more residents.

Megalopolis: a vast urban tract formed by the merging of conurbations and cities.

Metropolitan area: a large population concentration that typically includes an important city as well as the administrative areas bordering the city that are economically and socially integrated with it.

Millionaire city: a city with a population of 1 million or more people.

More economically developed country (MEDC): a relatively wealthy country in the later stages of development. The term includes all the countries in Europe and North America, along with Australia, New Zealand and Japan.

Newly industrialising country (NIC): a country that has recently undergone rapid development and is making the transition from LEDC to MEDC.

Pathway: a course or sequence of stages broadly followed by most countries as they progress in terms of development or urbanisation.

Polarisation: a widening of the gap between extremes — for example, between rich and poor.

Post-industrial: a term used to describe a set of changes and processes at work since the 1970s that have transformed cities, economies and societies. The process involves a shift of emphasis from manufacturing to services.

Primacy: where one city (usually a capital city) dominates completely the national urban hierarchy. It is significantly larger and more powerful than the second-ranking city.

Quality of life: the general state or condition of a population living in a given area.

Redevelopment: the demolition of old buildings to make way for new ones.

Regeneration: the revival of old, urban areas by redevelopment or improvement.

Re-imaging: changing the perceived image of an area by minimising the negative aspects and promoting the positives.

Renewal: see *redevelopment*.

Reurbanisation: the movement of people and economic activities back into city centres.

Rural dilution: the infiltration of urban lifestyles into rural areas.

Rural–urban continuum: the unbroken transition from predominantly rural to predominantly urban.

Shanty town: an area of slum housing built of salvaged materials and located either at the city margins or within the city on hazardous ground hitherto avoided by the built-up area.

Sink estate: an area of poor housing occupied by disadvantaged households.

Slum: an area of overcrowded, squalid housing and inadequate services occupied by poor households.

Social housing: housing for poorer families, usually provided by local authorities and rented by tenants.

Social malaise: the incidence of social problems such as crime, antisocial behaviour and ill health.

Social segregation: where particular groups of people live apart, either because they are forced to do so, or for economic or social reasons. Segregation is often based on wealth, ethnicity or age.

Squatter settlement: see *shanty town*.

Suburb: the built-up outer or peripheral parts of a town or city.

Suburban intensification: the process whereby suburbs become more urban as a result of increasing building densities, introducing non-residential activities and developing remaining greenfield sites.

Suburbanisation: the outward spread of the built-up area, often at lower densities compared with the older parts of the town or city. The decentralisation — of people first and then employment and services — is encouraged by transport improvements. The process is also encouraged by in-migrants settling around the urban fringe.

Sustainability: meeting the needs of people today without reducing the ability of future generations to meet their needs. Minimising the ecological footprint.

Top down: where national policies and decisions are handed down for local authorities to implement.

Transnational corporation (TNC): a large company operating in more than one country and typically involved in a range of economic activities.

Urban: countries differ in the way they define this term. Some define it as community or settlement with a population of 2000 or more, but there is more to the term than just population size. It is more correct to define it as relating to, or characteristic of, towns or cities, with their distinctive environments and ways of life.

Urban agglomeration: a contiguous urban area of 1 million people or more.

Urban environment: a built-up area of a town or city, together with the prevailing lifestyles, values and aspirations.

Urban fringe: the edge of a built-up area.

Urban hierarchy: the vertical classification of towns and cities according to a variable such as population size. This is best thought of as a pyramidal structure, with many towns as the base. Above these is a smaller number of cities, and above these, an even smaller number of regional centres. At the top of the pyramid there is a capital city.

Urbanisation: the growth in the proportion of a population living in urban areas.

Urbanism: the lifestyles, values, attitudes and behaviour that characterise people who live in towns and cities.

World city: see *global city*.

Part

A global view

What does 'urban' mean?

There is no universally accepted definition of the word 'urban'. However, most geographers agree that economic activity is important in distinguishing between urban and rural settlements, and between urban and rural people. An urban settlement survives mainly by activities such as manufacturing, mining and providing services. Rural settlements most commonly rely on farming, but fishing and forestry can also provide a livelihood. The transition over space between the countryside and the urban built-up area is known as the **rural–urban continuum**.

Urban settlements are generally larger than rural ones, but range enormously in size — from the small and simple to the huge and complex. A sequence of terms is used to cover this great range, starting with town and followed by city (formerly marked by its status in such fields as trade, government and religion). In some studies of **urbanisation**, the minimum size for a town is set at a population of 5000 and that for city at 100 000. The sequence carries on from city to metropolis, to **conurbation** and, finally, to **megalopolis**.

This sequence creates a vertical class system, known as the **urban hierarchy**, in the shape of a pyramid. At its base are many towns, above these lie a smaller number of cities, above these still fewer **metropolitan areas**, and so the structure tapers upwards. In some urban studies, the top three tiers of the urban hierarchy (metropolis, conurbation and megalopolis) are replaced by two tiers — **mega-city** and **global city**. Both are huge, with populations measured in tens of millions, but they differ in terms of their importance to the workings of the global economy; global cities being the lead players in that economy.

Two main types of urban hierarchy are recognised: primate and lognormal. In a primate hierarchy, the largest city in the country is well above the other leading cities. For example, in the urban hierarchy of Ethiopia, the capital city Addis Ababa has a population ten times that of its nearest rival, Dire Dawa (*Case study 4*). In contrast, China's hierarchy is lognormal. Shanghai is the largest city, but its population is less than 25% greater than that of the second city, the capital Beijing (*Case study 5*).

Therefore, urban settlements are distinguished by:
- their functions — the nature of employment; the range and provision of services
- their size — population and extent
- their high population densities

The interaction of these characteristics produces a distinctive, largely manmade environment. This gives rise to a particular way of life or lifestyle (sometimes referred to as **urbanism**); pace and stress are features of that lifestyle. Urban work differs from rural work not only in a functional sense — job opportunities are both more varied and greater in number. Success at work is converted into material possessions and progress up the social scale. Social interaction and upward mobility are not held back by old social structures — much importance is attached to what you do for a living and how much you earn. Improved services and more leisure opportunities are used to promote the belief that towns and cities offer a superior **quality of life**. However, higher rates of crime and other forms of **social malaise** suggest that people may behave differently in an **urban environment**. Self-interest tends to prevail over community care. Urban society is therefore highly polarised — great wealth and acute poverty exist almost side by side.

Urbanisation

The process of becoming more urban (the level of urbanisation) varies from place to place, as well as over time. With this in mind, we will try to take a 'snapshot' of the current state of play across the globe. This requires an international measure that allows countries and global regions to be compared. The percentage of the population living in urban settlements (often referred to as the **percentage urban**) is probably the most widely used indicator. However, this provides only a crude picture. Its value and accuracy are limited by the fact that:

- the definition of what constitutes an urban settlement — in particular, the minimum size threshold — varies from country to country
- towns and cities spread urban influences well beyond the limits of their built-up areas. For example, people living in country areas but working in a nearby town are probably better seen as belonging to the urban, rather than the rural population (see **'Counterurbanisation'**, pp. 23–25).

Case study 1 — THE URBAN WORLD

A snapshot

While the world's population is currently doubling every 40 years, the urban population is tripling. Within the next few years, more than half the world's population will be living in urban areas (Figure 2).

The degree or level of urbanisation differs considerably by region (Figure 3). North America has maintained its ranking over the last 50 years as the world's most urbanised region, followed by the other two 'more developed' regions of Europe and Oceania. However, the latter two regions have now been overtaken by Latin America. Among the **less economically developed countries (LEDCs)**, Latin

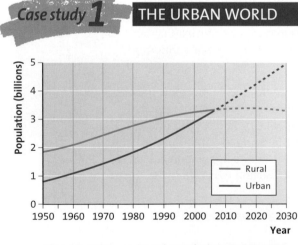

Figure 2 *Urban and rural populations, 1950–2030*

American countries have the highest proportion of their population living in urban areas. However, east and south Asia are likely to show the fastest growth rates over the next 30 years. Almost all future world population growth will be in towns and cities.

This case study makes three important points:
■ *Just over half the world's population is now urban.*
■ *The global regions show contrasting levels of urbanisation.*
■ *High rates of urbanisation mean that many LEDCs are rapidly catching up with the* **more economically developed countries (MEDCs)** *in percentage urban terms.*

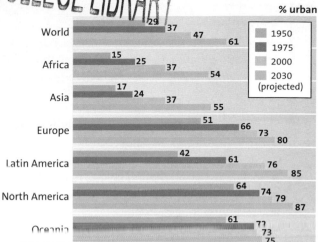

Figure 3 *Population living in urban areas by region, 1950–2030*

Question

Study Figure 4 and identify the main features of the global distribution of urban population.

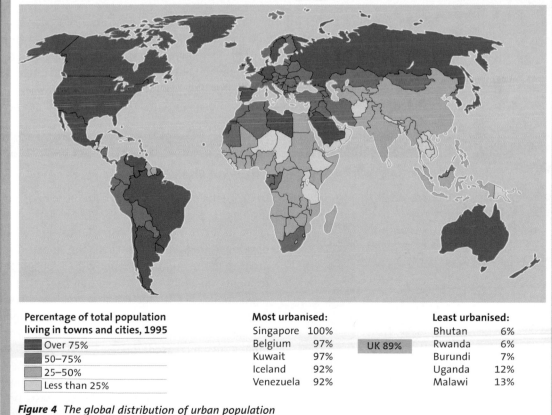

Percentage of total population living in towns and cities, 1995
- Over 75%
- 50–75%
- 25–50%
- Less than 25%

Most urbanised:
Singapore	100%
Belgium	97%
Kuwait	97%
Iceland	92%
Venezuela	92%

UK 89%

Least urbanised:
Bhutan	6%
Rwanda	6%
Burundi	7%
Uganda	12%
Malawi	13%

Figure 4 *The global distribution of urban population*

Guidance

One of the skills tested in many A-level geography examinations is the ability to describe features on a distribution map. The global distribution of urban population is a good map to focus on. Your strategy should be to identify any outstanding features. A useful tip is to concentrate on the 'extreme' areas, such as those showing the highest and lowest values. Deal with each of these in turn, using bullet point notes. Pointing out the exceptions to the generalisations you have made should be kept to a minimum. Finish by focusing on the middle-value areas.

- High values (over 75%) in North America, much of South America, Australasia and in the extensive Russian Republic.
- Scattered high values in Europe, north Africa and the Middle East.
- Lowest values (less than 25%) scattered in parts of Africa and south Asia.
- Median values (25–75%) in much of Europe, Africa and Asia, Central America and South America.
- Brief reference might be made to the possible 'exaggerated' importance of urbanisation in Greenland and Russia, where vast areas are unpopulated.
- Remember that the question has not asked you to explain, just to identify (describe).

Urban agglomerations

The rising tide of global urbanisation is a spectacular feature of the last 50 years. Equally impressive is the rapid concentration of this growing urban population in large cities. It may be more accurate to refer to them as **urban agglomerations** rather than cities, as many are formed when a major city expands and engulfs other settlements. The term **mega-city** is currently popular, but officially it should be reserved for the very largest, namely those containing at least 10 million people.

Case study 2 — **MEASURING IN MILLIONS**

Cities on the rise

By 2015, the number of cities with more than 1 million residents is projected to be about 564, a rise from 195 cities in 1975 (Figure 5). Asia, Africa and other less developed regions have seen the most dramatic increase in the number of cities with 1 million or more residents, as well as in the proportion of the total population concentrated in these 'million-plus' cities.

In more developed countries, including Japan and the USA, the number of cities with at least 1 million inhabitants increased modestly between 1975 and 1995, rising from 85 to 114. This number is expected to reach 138 by 2015, the latest year for which individual city projections are available. By contrast, between 1975 and 1995 the number of million-plus cities in less developed countries soared from 110 to 250, and it is expected to surpass 425 by 2015.

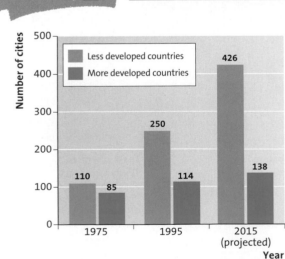

Figure 5 *Number of cities with 1 million or more residents, 1975–2015*

This scaling up of urban growth is now reaching new heights with the emergence of urban agglomerations with populations of 5 million or more (Figure 6). In 1950, there were eight agglomerations, today there are 40, and by 2015 it is estimated there will be 56.

Size of urban population

○ 5 million and over since 1950
● 5 million and over since 2000
◉ 5 million and over in 2015 (projected)

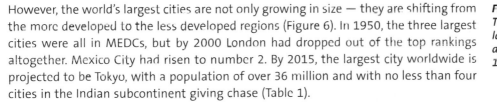

However, the world's largest cities are not only growing in size — they are shifting from the more developed to the less developed regions (Figure 6). In 1950, the three largest cities were all in MEDCs, but by 2000 London had dropped out of the top rankings altogether. Mexico City had risen to number 2. By 2015, the largest city worldwide is projected to be Tokyo, with a population of over 36 million and with no less than four cities in the Indian subcontinent giving chase (Table 1).

Figure 6
The world's largest urban agglomerations, 1950–2015

Table 1 *The ten largest urban agglomerations, 1950–2015, using UN data 2003*

Rank	1950		2000		2015 (projected)	
1	New York	12.3	Tokyo	35.0	Tokyo	36.2
2	London	8.7	Mexico City	18.7	Mumbai	22.6
3	Tokyo	6.9	New York	18.3	New Delhi	20.9
4	Paris	5.4	São Paulo	17.9	Mexico City	20.6
5	Moscow	5.4	Mumbai	17.4	São Paulo	20.0
6	Shanghai	5.3	New Delhi	14.1	New York	19.7
7	Essen	5.3	Kolkata	13.8	Dhaka	17.9
8	Buenos Aires	5.0	Buenos Aires	13.0	Jakarta	17.5
9	Chicago	4.9	Shanghai	12.8	Lagos	17.0
10	Kolkata	4.4	Jakarta	12.3	Kolkata	16.8

There are six main factors encouraging this large-scale urban agglomeration (Figure 7):

■ **Communication economies**: the benefits that people and businesses reap as a result of being close together, such as easy face-to-face contact, low transport costs, sharing modern **infrastructure** and specialist services costs.

- **Cumulative causation**: by capitalising on some initial advantage, the successful city becomes even more successful — a multiplier effect kicks in.
- **Population growth**: the basic input that comes from two sources — high rates of natural increase and positive net migration.
- **Mass-gravity effect**: the momentum that causes large objects to become even larger.
- **Status-symbol syndrome**: the 'buzz' felt by people and businesses because they perceive that, by being part of a large city, they are somehow close to the 'cool' centre of things — there is kudos and 'street cred'.
- **Development**: the scaling up of urban growth is only possible if supported by economic development and advances in technology.

Figure 7 Factors encouraging large-scale urban agglomeration

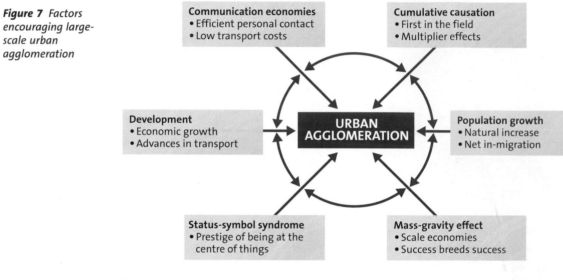

This case study emphasises two points:
- the relentless scaling up of cities
- the rapid spread of mega-cities in LEDCs

Using case studies 2

Question

(a) Describe the rise and global spread of 5-million cities between 1950 and 2000, as shown in Figure 6.

(b) Suggest reasons for this global spread.

Guidance

(a) ■ Two factors need to be taken into account: the increase in number from 8 to 40, and the change in distribution.
  ■ The distribution change has three aspects: more such cities in the original regions of Europe, the Americas and east Asia; the proliferation in Asia (China and Indian subcontinent); and the spread to Africa.

(b) ■ Rising levels of development, particularly industrialisation and the growth of the tertiary sector.
  ■ High rates of natural increase and rural–urban migration.
  ■ Now make use of the other factors given in Figure 7.

A class apart

Mega-cities are urban areas with populations greater than 10 million. There are currently 24 of them. Global cities, on the other hand, can be of any size. At present, there are 30 such cities. They all have populations over 1 million and nine of them are mega-cities, with London being the exception.

What distinguishes a global city from a mega-city? Global cities are recognised worldwide and are unchallenged as seats of prestige, status, power and influence. As Figure 8 suggests, these star qualities have their roots in a number of different fields, the most important of which is economic. All global cities are key hubs in the emerging global economy. Above all else, globalisation has encouraged the emergence of this star category in the world's urban hierarchy. Besides possessing certain key economic attributes, the aspiring global city needs to satisfy a range of demographic, technological, cultural and political requirements, but not necessarily all of those identified in Figure 8.

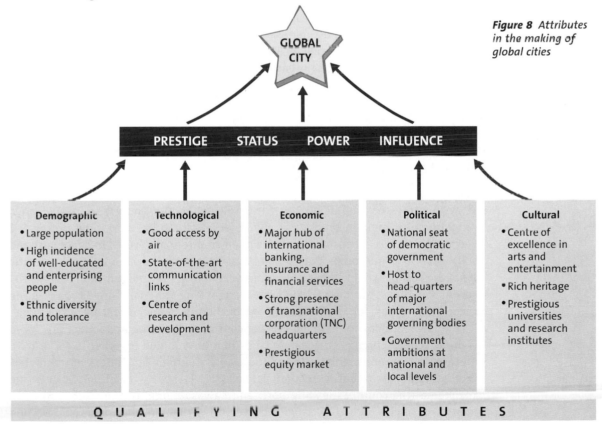

Figure 8 Attributes in the making of global cities

There are two types of global city. The first is the elite top grouping of three: London, New York and Tokyo. Each of these has an array of lesser world cities within its immediate orbit — there are ten for London, nine for New York and five for Tokyo. The second type includes the four global cities of the Southern Hemisphere (Buenos Aires, Rio de Janeiro, São Paulo and Sydney). These appear to be out on a limb, but efficient transport and communications are able to overcome their apparent isolation. A little under half of all global cities are not capital cities. This raises the interesting question of what has provided the power to raise these cities — such as Frankfurt and Milan in Europe and Los Angeles and Toronto in North America — to stardom? Other interesting questions include the following:

■ Are there many cities waiting in the wings to become world cities? Consider Shanghai, for example.
■ Which city of the Indian subcontinent will be its first global city: Mumbai, Delhi, Kolkata or Dhaka?
■ When will Africa produce a global city, and which city might it be? Is Johannesburg the most likely candidate?
■ Is there a limit to the number of cities that can belong to this exclusive club?

The following examples of Dhaka and Tokyo may help to illustrate the difference between a mega-city and a global city.

Dhaka

When Bangladesh became an independent state in 1971, Dhaka became the capital. At that time, the level of urbanisation stood at less than 10%. Today, it is roughly 25%. The latest UN population estimates suggest that, by 2015, the national population will have exploded to 150 million, and that of Dhaka will reach nearly 18 million. It will then rank seventh largest urban agglomeration in the world (Table 1). However, Bangladesh is one of the world's poorest and most hazard-prone countries. Dhaka is a chaotic city, facing a host of problems that include poverty and poor housing, congested roads and frequent flooding. Its involvement in globalisation is limited and relates mainly to what outsiders see as the country's main resource: cheap labour. Transnational corporations (TNCs) are setting up branch plants there and many are involved in manufacturing clothing.

Tokyo

Tokyo became the capital of Japan in 1867, when the country embarked on a programme of modernisation and industrialisation. For a short time during the Second World War, it became the centre of an empire that covered much of east and southeast Asia. Although that empire was lost and the country defeated, the Japanese economy made a miraculous recovery during the 1960s. It soon became the second wealthiest economy in the world — a ranking still maintained today. During these boom times, Tokyo emerged as one of the top financial centres in the world. It is renowned today for its high levels of consumer spending and modernity. Tokyo is in fact one of three once-separate cities that have joined together to form Tokyo Metropolis. Currently, this is the world's largest urban agglomeration and it is forecast to maintain that rating until at least 2015 (Table 1). The other two cities are Yokohama (the country's leading port) and Kawasaki (one of the main centres of manufacturing). Given this economic diversity within the metropolitan area and Japan's economic status, it is hardly surprising that Tokyo should be recognised as a super-elite global city.

This case study shows that there is much more to a city becoming a global city than just achieving a certain size. Of all the many different attributes that distinguish a global city, it is those of an economic nature that are paramount.

Question

What distinguishes a global city from a mega-city?

Guidance

You could base your answer on the comparison of Dhaka and Tokyo in *Case study 3*. You may need to do a little research of your own to find out what Tokyo has that Dhaka lacks.

Question

Read the following newspaper extract and identify the main ways in which mega-cities are changing today.

What is happening to our mega-cities?

'A city is a living thing. It has a complex metabolism, a voracious appetite and very poor eyesight.' So says Herbert Girardet, famous for his work on city metabolisms and ecological footprints.

The phenomenon of mega-cities is not new, but the problems faced by the Greeks in the original Megalopolis (population was probably only 40000) do not compare with those suffered by the sprawling giants of the twenty-first century. The global total of mega-cities continues to rise; so too do the problems they face.

It might be thought that if a mega-city stops growing, then it might be better managed. But cities rely on the injection of new ideas and labour that come with the arrival of new people. The overriding problem is that the size and form of mega-cities, with their central focus on the CBD, have become so congested and unpleasant to live in that the 'flight out' of people, businesses and their associated wealth has started to reduce the vibrancy of such cities.

The original nuclei of Kolkata, Mexico City and São Paulo have all shown recent signs of a slackening rate of population growth. The predictions of monstrous growth by the twenty-first century have simply not materialised. So what has happened? In India, Kolkata has spawned a series of new urban centres in West Bengal. Mexico City is losing ground to the country's middle-order cities.

In the case of London, it could well be argued that, while its population has slipped back to 8 million, it is now simply part of a megalopolis, stretching from Milton Keynes to Dover, from Cambridge to Basingstoke and soon to incorporate the Thames Gateway.

In China, Hong Kong is part of the rapidly expanding Pearl River delta megalopolis. But the fastest growing urban area in the world is the Yangtze delta, which is home to a crescent-shaped urban corridor of 16 mega-cities, including Shanghai. An estimated 75 million people are already living here on only about 1% of China's total land area. This megalopolis dominates China's production and exports, and is the target of almost half the foreign investment being made in the country. Located in the middle reaches of the Yangtze and with half a million people arriving annually in search of a better life, Chongqing is the world's fastest-growing and biggest municipality. It has 31 million residents; more people than the total population of Iraq or Malaysia. Its congestion, air pollution and refuse problems are, however, phenomenal too.

The precedent of several urban agglomerations coalescing to create a megalopolis was set by the so-called BosWash megalopolis along the eastern seaboard of the USA and by Japan's Tokaido megalopolis. With a population of 70 million people, the latter is the largest and densest urban sprawl in the world. However, so far it has the finance and technology to ensure the general quality of life is maintained despite continuing growth. Can the same be said for São Paulo's 'golden urban triangle' or the Yangtze delta megalopolis?

Guidance

You should be able to identify at least four different ways in which mega-cities are changing. One is that they are still increasing in number. Now think in terms of size, trends and quality of life. Look at *Case studies 2, 5* and *6*.

Websites: www.megacities.uni-koeln.de/frame.htm
www.megacities.uni-koeln.de/internet/start.htm
www.china.org.cn/english/2004/Sep/108171.htm
http://travel.guardian.co.uk/countries/story/0,1731346,00.html

The urbanisation pathway: early phases

Countries move along a common **urbanisation pathway**, as shown in Figure 1(A) (p. viii). This pathway is sometimes referred to as the urbanisation cycle or the urbanisation curve. The first term might give the impression that the series of changes occurs more than once. However, the only repetition is spatial, in that different parts of the world are experiencing the same general sequence. They may do so at different times and speeds, but they do so only once.

The aim in parts 2 and 3 is to illustrate the particular urban characteristics of a sample of countries, each at a different stage along the urbanisation pathway. Before doing so, however, it is important to be clear about the nature of the urbanisation process.

Urbanisation and development

Urbanisation is a process of change whereby places and people become increasingly urban. It is an essential part of the broader development process; development is the force that drives urbanisation. It is a multi-strand process and five changes are particularly important (Figure 9):

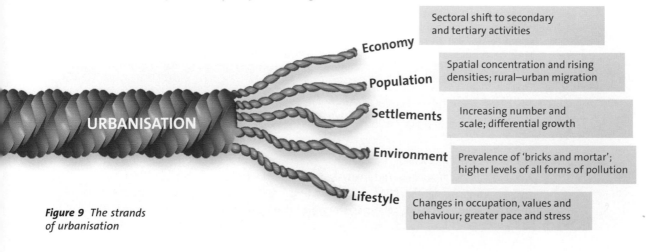

Figure 9 The strands of urbanisation

Economy — Sectoral shift to secondary and tertiary activities

Population — Spatial concentration and rising densities; rural–urban migration

Settlements — Increasing number and scale; differential growth

Environment — Prevalence of 'bricks and mortar'; higher levels of all forms of pollution

Lifestyle — Changes in occupation, values and behaviour; greater pace and stress

- **Change in the economy:** the emphasis shifts from farming and the primary sector to manufacturing and the provision of services.
- **Change in the distribution of population:** the concentration of people and their non-agricultural activities at favoured locations leads to the birth and growth of towns and cities. A vital contributor is rural–urban migration.
- **Change in the size and character of settlements:** some settlements, particularly those at favoured locations, grow more quickly than others. Differential growth sees some villages grow into towns, some towns into cities and some cities into urban agglomerations.
- **Change in environment:** the built-up area is a distinctive man-made environment with raised levels of pollution.
- **Change in lifestyle:** this applies to people helping to swell the populations of towns and cities. It is not just a change in occupation, it is a change in values, behaviour and social institutions. There is greater pace and stress.

Having identified these five main strands of change, we can see that urbanisation has other important features:

- It is initially **centripetal** or centralising in character because towns and cities act as magnets, drawing in people and activities. Later, the direction is reversed as **decentralisation** becomes dominant — that is, centrifugal forces begin to operate.
- It is a spatial process in that it causes places and environment to change.
- It is also a spatial process in that, at any one time, some countries or regions will achieve higher levels of urbanisation than others.
- It is a process that can operate at a whole range of speeds. In some places, it may be slow and almost imperceptible, while in others it can be fast and frightening, particularly when governments deliberately promote cities. The speed is governed not only by the overall speed of development in the country or region, but by the global economy.

In this chapter, the focus is on the first three of the five phases along the urbanisation pathway (Figure 1(A), p. viii):

- The **low income** or **rural society** phase: although some towns and cities may exist, the level of urbanisation is low and rises only imperceptibly.
- The **lower-middle income** or **take-off** phase: the rate of urbanisation accelerates.
- The **upper-middle income** or **drive-to-maturity** phase: the rate of urbanisation becomes steady, but the overall level of urbanisation continues to rise impressively.

All three phases have one thing in common: the forces of **centralisation** prevail.

ETHIOPIA

Case study **4**

A late starter

With a population of around 67 million, Ethiopia is one of Africa's largest and most populated countries. It also one of the world's poorest, with 60% of its population living below the poverty line. With per capita **gross national income** (GNI) at US$90, it is almost bottom in the global rankings. For centuries, economic development and urbanisation have been hampered by:

- tribal conflicts
- religious wars
- colonial invasions
- recurrent famines
- a recent period of socialist misrule

However, since 1991, the situation has changed radically as Ethiopia has begun to make the transition from a socialist country to a market economy. It is becoming increasingly exposed to the forces of **economic globalisation**. During the 1990s, the urban percentage of the population increased from 12% to 15% in only 10 years. The current rate of urbanisation is high, at 4.6% per annum. This compares with 2.6% in China, 1.9% in Mexico, 1.4% in the USA and 0.3% in the UK.

Rank	City	Population (2004)
1	Addis Ababa	2 763 500
2	Dire Dawa	254 500
3	Nazret	176 800
4	Gondar	147 900
5	Mekele	133 500
6	Bahir Dar	131 800
7	Dese	126 300
8	Jimma	117 600
9	Harer	99 600
10	Awasa	99 100
11	Debre Zeyit	98 700
12	Shashemene	88 500

Table 2 The primacy of Addis Ababa

Addis Ababa

The outstanding characteristic of Ethiopia's urbanisation is the emergence of Addis Ababa, the capital, as a highly primate city. It was only founded in 1881. Now with a population of almost 2.8 million, it is more than ten times larger than the second city, Dire Dawa (Table 2). It is predicted that by 2015, it will become the fourth largest city in Africa. Historical events that have helped to bring about this **primacy** include:

- the decision taken in 1896 to establish the residence of the Ethiopian royal family in the city
- focusing foreign investment from Europe, the Middle East and India on the city during the late nineteenth and early twentieth centuries
- completing a railway from Addis to Djibouti in 1917, which gave the city a direct link to the coast

The first of these events made Addis Ababa the seat of political power, while the latter two strengthened its position as the economic capital of the country. More recently, Addis Ababa's primacy has been intensified by:

- the high rate of natural increase in its population
- large volumes of in-migration, from rural areas and from other towns and cities
- the convergence on the city of ex-soldiers and redundant civil servants
- the growth of international tourism that currently focuses on the city
- the choice of the city as the location for the secretariats of both the Organization of African Unity (OAU) and the United Nations Economic Commission for Africa (UNECA)

The first two factors are readily evident in the leading cities of most LEDCs. In Addis Ababa's case, in-migration from rural areas has been increased by natural disasters, famines and uncertainties about rural land tenure. Its population almost doubled between 1984 and 2004.

Ethiopia's emerging urban hierarchy comprises 12 cities (Table 2). The pace of urbanisation in Ethiopia today might be even faster, were it not for the long-running border dispute with Eritrea (once part of Ethiopia). Much of Ethiopia's limited gross national income (GNI) is still being used to pay for military defence, rather than to support further development and urbanisation.

This case study illustrates some of the following characteristics typical of a country moving through the first phase of the urbanisation pathway:

- *the general link between development and urbanisation*
- *the important role of rural–urban migration*
- *the convergence of much of this migration on the largest city, thereby increasing its primacy*
- *the debilitating impact of primacy on other towns and cities*
- *the boost given to urbanisation when things go wrong in rural areas*
- *the prevailing strength of the forces of centralisation*

---

4 Using case studies

Question

(a) What is meant by a 'primate city'?

(b) With reference to examples, suggest reasons for the rise of primate cities.

Guidance

(a) A primate city is not only the largest in a country — it is much more than twice the size of the second largest city and completely dominates the national urban hierarchy.

(b) The following reasons may be extracted from *Case studies 3, 4* and *6*:

- the function of the capital city
- the accessibility of the city from all corners of the country

- the compactness or small extent of the country
- some initial or comparative advantage (e.g. coastal location or natural resources)
- the lack of political will to limit its own success and encourage urban growth in lower-order cities instead

---

CHINA

Case study **5**

Looking to the future

In terms of the development pathway, China is classified as a 'lower-middle-income' country. With per capita gross national income (GNI) at US$1100, it is 133rd in the global rankings. The country has a long urban history, dating back to the second century BC. It was not until the last century, however, that China really began to proceed along the urbanisation pathway.

In 1950, the percentage urban figure was 11. By 2000, it had almost trebled to just over 30. That figure might have been even higher had the communist government not limited rural–urban migration, as happened during the Cultural Revolution of the late 1960s. China's urban population is currently growing at a rate of 2.6% a year — one of the fastest rates in the world. This compares with 0.8% for India — another large and fast developing nation. A policy of decentralisation from the leading cities — Beijing, Shanghai, Guangzhou and Tianjin — has already led to the creation of many new cities. However, because the established coastal cities are much better placed for trade and are more attractive to foreign **inward investment**, serious inequalities are beginning to emerge within the urban system.

The Chinese authorities see promoting urbanisation as the best way to:
- modernise and strengthen the economy
- tackle poverty
- reduce the gap between the backward western regions and the wealthier east of the country

Chinese economists believe that there is a strong positive correlation between urbanisation and economic development. The official view is that a country in which most of the population is in poor or remote villages will not be a modern and developed nation. In many rural communities, farmers are forced to supplement their small incomes with part-time labour or small trading. They see the wealth gap between them and nearby towns widening. Cheap labour is needed desperately in the urban factories that are beginning to turn China into 'the workshop of the world'. If there is the promise of work and a regular wage, rural people do not require much persuasion to move.

The Chinese government has recently decided that between 300 million and 500 million people should move from rural to urban areas over the next two decades. This could become the biggest migration in human history. Over the same period, it states that some 400 new cities and 10 000 new towns should be created, mainly to accommodate people leaving their farms. Many of these will be located in rural areas that are currently major source regions of migrants. China already has over 650 cities (Table 3). The overall result of these plans is that the country's urban population will rise to around 800 million by 2020, accounting for between 55% and 60% of the national population.

Size (millions of people)	Number of cities
>2	13
1–2	27
0.5–1	53
0.2–0.5	218
<0.2	352

Table 3 Chinese cities by size, 2000

At the same time as these new cities and towns are being built, everything possible is being done to promote Beijing and Shanghai as global cities.

There are, however, potential dangers in this programme of promoting rapid urbanisation. These include:
- rural areas becoming impoverished by the loss of people of working age
- not meeting the infrastructural needs (housing, schools etc.) of the rapidly expanding urban population
- negative environmental impacts, especially air pollution
- rising social unrest in the cities, as the poor incomers settle next to wealthy urban people

China has been described as a country of 'two systems and four societies'. Its rural and urban residents live in two separate 'systems'. Urban people live in a system that increasingly recognises market forces and grants them some freedom in matters of housing, education, public services and employment. In contrast, rural people live in a more rigid system that still has many of the hallmarks of hard-line communism of the Maoist variety. The four 'societies' are farming, manufacturing, services and knowledge. They are ranked in this ascending order in terms of increasing wealth and influence. The low ranking of agriculture underlies the growing income divide between rural and urban populations. Only 5% of China's workforce is employed in what is called the 'knowledge society' — technology, education, health, finance, business and the civil service. This society may be small in number, but its wealth and influence are disproportionately large.

Shanghai

With a population of 16.7 million in 2000, Shanghai is China's largest city. It was established more than 700 years ago on the coast to the south of the Yangtze delta. By the beginning of the nineteenth century, it had a population of over half a million people and was a thriving commercial centre, largely due to its proximity to the cotton-growing regions of China. Its prosperity was given a boost in 1842 when it became one of the first Chinese ports opened by treaty to foreign trade.

Today, Shanghai is a major port as well as an important centre for heavy industry. Recently, e-businesses and e-sales have been expanding rapidly. Over half of the Fortune 500 companies (the most profitable companies in the world) and nearly 18 000 other foreign enterprises operate in the city. It is also home to nearly 200 research institutions. Shanghai has become the first provincial area in China to reach, in terms of **gross domestic product (GDP)** per capita, the level of a middle-income country. As one of globalisation's 'hot spots', it has been chosen to spearhead China's move to become a more market oriented economy.

www.flickr.com/photos/pmorgan

Figure 10
Downtown Shanghai

Unlike many other LEDC cities, Shanghai's growth, since communism took over in 1949, has been firmly in the hands of state and city planners. However, in that time there has been a fundamental reversal of strategy, from centralisation to decentralisation. Current policies focus Shanghai's new growth on seven satellite towns being built around the city (Figure 29). See also *Case studies 18, 29 and 34.*

This case study illustrates:

- *how the government of the world's largest nation firmly believes in the critical link between urbanisation and economic development*
- *how, in a socialist economy, government can be much more pro-active in the promotion of urbanisation and of individual cities than in a capitalist economy*
- *the importance of matching population growth and rising consumer demands with the provision of housing, services and amenities*
- *that the leading or largest city of a country is not always its capital*

MEXICO

The meteoric rise of a mega-city

With per capita gross national income (GNI) currently standing at US$6330, Mexico is one of the 'upper-middle-income' economies. Development and urbanisation over the last 50 years have transformed the country. Mexico now ranks as one of Latin America's most urbanised countries.

At the beginning of the twentieth century, only 10% of Mexico's population lived in urban settlements. Over the next four decades, this figure doubled to 20%. In contrast with earlier and rather turbulent times, the second half of the twentieth century was a

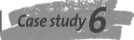

period of relative political stability. This created the correct environment for a surge in economic development and the take-off in urbanisation that now puts the percentage urban figure at over 75% (Figure 11). Four economic factors played an important role in that development:

■ Mexico's rich natural resource base of minerals and oil
■ the cheapness of its labour — this and the natural resources are attractive to TNCs
■ the government's decision to promote import substitution (i.e. manufacturing goods that were formerly imported)
■ the boom in the service sector, which now generates almost 70% of Mexico's GDP

Figure 11 Mexico: the growth of population, urbanisation and Mexico City, 1950–2000

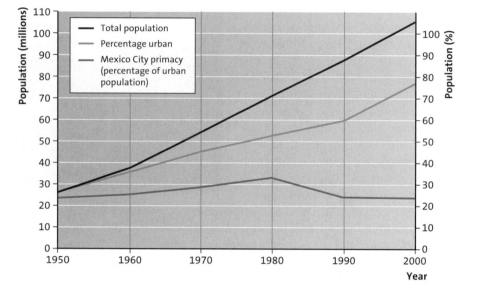

Mexico's urbanisation is a 'tale of one city', with Mexico City emerging as a primate city. The metropolitan area has a population over five times larger than Guadalajara, the second city, and six times larger than Monterrey, the third city.

Mexico City

Mexico City is one of the oldest continuously inhabited sites in the Americas. It is located within the central Mexican plateau, on the former bed of Lake Texcoco, and is surrounded by mountains over 5000 m high.

Mexico City's centre (the Zocalo) was an Aztec capital. By the early 1500s, the city was the hub of a powerful commercial and military empire stretching from Texas in the USA to Honduras in Central America. It included magnificent palaces, temples and markets, and was inhabited by around 200 000 people.

In 1519, Spain conquered the city, turning it into a colonial capital. So began 300 years of foreign rule. In 1821, Mexico gained its independence, but it was not until 100 years later that Mexico City really began to develop as a centre of manufacturing. Its population in 1821 was only a little larger than it had been in 1519. During the twentieth century, its rate of growth was far faster than anything experienced in MEDCs. In 1940, the population was around 1.5 million. By 1970, it was 9 million and in 2000, 18 million, making it the world's second largest urban agglomeration (Table 1, p. 5).

The primacy of Mexico City has been the outcome of:

- the concentration of much of Mexico's new manufacturing and services there
- high levels of internal rural–urban migration, swollen by people wishing to escape the unrest and unemployment that prevailed in many other parts of the country
- high rates of natural increase in both the established and the in-comer components of the city's population
- significant amounts of immigration from Argentina, Brazil, Cuba, Spain, France, Italy and Japan

Of the 20 million people living in the metropolitan area of Mexico City today, 40% are migrants and it is estimated that around 2000 more people settle in the city each day. Mexico City now accounts for 49% of the country's manufacturing and 68% of its financial services. The changing distribution of population within the growing metropolitan area is shown in Figure 12.

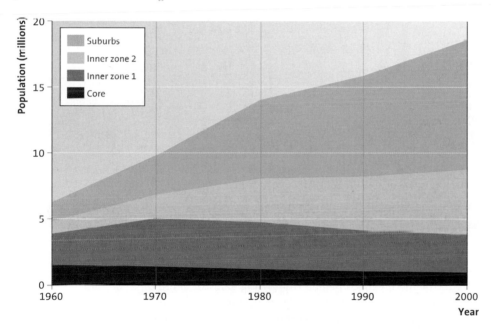

Figure 12
The changing distribution of population within the Mexico City metropolitan area, 1960–2000

There are now signs that the rate of growth is beginning to ease. It is estimated that by 2015 the city's population may be less than 25 million and that its global ranking will fall to six. There are also some signs of a shift taking place within the urban hierarchy. During the last two decades, the fastest rates of growth occurred in middle-order cities, with populations ranging from 100 000 to 1 million. This may be an early indication that Mexico is beginning to approach that critical point along the urbanisation pathway at which centralisation begins to give way to decentralisation. Figure 11 suggests that Mexico City's primacy passed its peak in 1980. The 1985 earthquake that hit the city and the economic crisis of the 1980s may have been responsible for this.

Like many world mega-cities, Mexico City has all the problems associated with unplanned and haphazard development. These include severe pollution, inadequate housing and services, high population densities (in some parts reaching 500 people per hectare) and a generally poor quality of life for most residents. Some 60% of the city's population live in settlements that were originally illegal. See also *Case study 13*.

This case study illustrates the following points:

- *A remarkable pace of urbanisation can be achieved by development, particularly when it is given a free rein.*
- *Urbanisation can become focused to an incredible degree in one all-dominant, primate mega-city.*

5

Using case studies

Question

What information about urbanisation in Mexico can be drawn from Figure 11?

Guidance

- Start by focusing on the 'percentage urban' curve. This shows a steady rate of increasing urbanisation between 1950 and 1990. There is a noticeable quickening of this rate in the 1990s.
- Compare this curve with the curve for 'total population'. Between 1960 and 1990, the rate of increase was less than that for total population. There was therefore a considerable amount of population growth in rural areas. Before 1960 and after 1990, the rates of increase were more or less the same.

- The curve for Mexico City shows that there has been a slight slip in its primacy (in terms of its percentage share of the total urban population). Between 1950 and 1980, there was a slight increase in this percentage figure. However, since then, it has slipped. This suggests that either the pull of Mexico City is now being weakened by the growth of rival cities (counterurbanisation), or there has been a reduction in the volume of rural–urban migration.

Case study 7 — IRAQ

A case of slipping back

During the 1970s, Iraq first achieved the status of a middle-income country, largely because of the revenues gained from rising sales of crude oil in an oil-thirsty world. Initially, the money was spent wisely on improving Iraq's physical and social infrastructure. The health and education systems of the country soon became among the best in the oil-rich Middle East. This new-found wealth prompted a take-off in urbanisation. Between 1950 and 1970, the percentage urban figure rose from 35 to 56, and by 1990 it had reached 74. Helping to boost these figures were a high rate of natural increase and the physical character of much of the country (desert and high-risk flood plain). The rapidly rising population could only be sustained by non-agricultural activities and by being accommodated in towns and cities.

In terms of development and urbanisation, Iraq today presents a different picture compared with 30 years ago. It has been crippled by:

- the corrupt leadership of Saddam Hussein, whose dictatorship lasted from 1979 to 2003 — he and his regime misused the country's funds to buy and develop armaments, in support of military campaigns, both within and outside the country
- a succession of disastrous wars with Iran, Kuwait, the USA and its allies
- internal conflicts, particularly between the ruling Sunnis on the one hand and the Shi'ah Muslims and Kurds on the other

- severe sanctions imposed by the international community after Iraq's defeat in 1991, particularly on its export of oil
- the recent anarchy following the removal of Saddam Hussein

Many people have fled the larger cities to escape persecution, aerial bombing and the terror campaigns of insurgents, so it is hardly surprising that the percentage urban figure has fallen back to 70. Poor housing and poverty prevail in the cities — infrastructure and services are wholly inadequate. Nationally, per capita income and living standards are now well below their pre-1990 levels. Iraq is no longer a middle-income country.

Baghdad

The urban hierarchy of Iraq is dominated by the capital city, Baghdad. With a population of 5.6 million, it is roughly four times bigger than the second and third cities, Mosul and Basra. It accounts for about a quarter of the Iraqi population.

Baghdad is one of the oldest and most celebrated cities of the Arab world. It was established in the eighth century, on the banks of the River Tigris. In the middle of ancient Mesopotamia, the city ruled over much of central Asia, Arabia, north Africa, and today's Spain. For centuries, it was a vibrant, multi-ethnic metropolis. Perhaps it was an early example of a global city. The city was home to large Jewish and Christian populations, as well as Kurdish, Turcoman and Persian minorities. But that has now changed. Most of the minorities have left, in response to pressures applied by Saddam Hussein and the ruling Sunni Muslims.

A major stimulus to the growth of the city in the 1950s came from Iraq's rapidly expanding oil industry. Initially, the money was well spent, for example on a huge flood protection scheme involving the diversion of the River Tigris. The oil revenues paid for paved streets, sewerage systems and street lights. In just 10 years, the city's population doubled from 500 000 to 1 million. This spectacular growth resulted from a great influx of rural migrants, drawn by employment opportunities. Most of these migrants settled in shanty towns called 'sarifahs', where housing was constructed from mud, tin cans or reeds. Sadly, very little money was spent planning and controlling this growth, so urban sprawl soon became a problem. The edge of the built-up area is now well over 15 km from the city centre.

In Baghdad, the narrow streets have persisted as a feature of the urban landscape. With their meagre opening to the sky, they stay relatively cool in summer. On the other hand, the maze of narrow streets continues to adversely affect other aspects of urban life. Housing is congested and inadequately serviced, and the use of cars is greatly restricted. Despite several clearance campaigns, both before and after Saddam Hussein came to power, the shanty areas continue to scar the built-up area. Baghdad today also bears new scars — those of damage and disruption caused by the recent war that toppled Saddam Hussein and by the serious civil unrest brought about by insurgents intent on destabilising the country (Figure 13).

Many outside international agencies are now working on a massive city recon-struction programme. This involves the repair of war-damaged areas and making good the several decades of under-investment in Baghdad's infrastructure. The immediate challenges are water supply, sewage treatment, electricity generation and restoring normality to city life. It is estimated that this reconstruction programme will cost at least US$413 million. Therefore, the sooner Iraq's oil revenues are restored, the better.

Figure 13 *War-torn Baghdad*

TopFoto

The plight of Iraq today is not a happy one. Its once proud flagship city is bruised and battered. Baghdad is no longer a global city.

This case study makes two points:
- *Movement along the urbanisation pathway is not always forwards.*
- *Corrupt governments, war and terrorism can lead to retreat on that pathway and the demise of once-proud and influential cities.*

Using case studies 6

Question
With reference to examples, suggest reasons why cities may stagnate or decline.

Guidance
- Baghdad provides a good example to start with, because of war and civil unrest.
- Find examples of cities affected by the following: resource exhaustion; natural hazards; economic relocation; corrupt government; physical processes.
- What other causes of stagnation or decline can you think of?

Correlations

Case studies 4–7 looked at four countries at different stages along the development and urbanisation pathways. Two positive correlations emerge:
■ between the level of development and the level of urbanisation
■ between the rate of development and the rate of urbanisation

These correlations exist because development drives urbanisation. Figure 14 shows the different aspects of development that help to create these correlations with urbanisation. Three spatial scales are involved. The emergence of the global economy clearly creates a range of opportunities for towns and cities to prosper. For example, the transnational corporations' endless search for accessible, cheap labour stimulates urbanisation in some poorer parts of the world. At a national level, the growth of the secondary and tertiary sectors at the expense of the primary sector may be the shift that most encourages urbanisation. Once this shift starts, local factors come into play to determine where the urbanisation process will flourish most. **Comparative advantage** determines which urban settlements prosper.

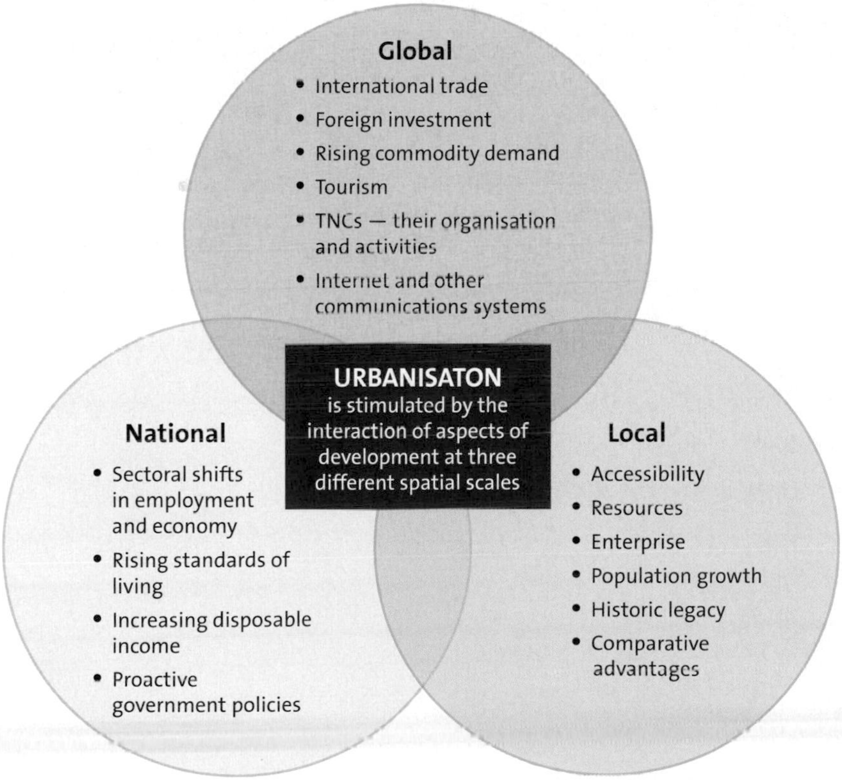

Global
• International trade
• Foreign investment
• Rising commodity demand
• Tourism
• TNCs — their organisation and activities
• Internet and other communications systems

URBANISATON
is stimulated by the interaction of aspects of development at three different spatial scales

National
• Sectoral shifts in employment and economy
• Rising standards of living
• Increasing disposable income
• Proactive government policies

Local
• Accessibility
• Resources
• Enterprise
• Population growth
• Historic legacy
• Comparative advantages

Figure 14 Aspects of development that encourage urbanisation

Case studies 4–7 demonstrate that agglomeration is the dominant process of urbanisation during the early phases of the pathway. Furthermore, unless there is management and control, agglomeration can lead to primate cities emerging. These act as a drain on other parts of the country, and create many internal problems. Some of these problems are examined in part 5.

Question

With reference to LEDC examples, examine the links between urbanisation and development.

Guidance

The essay should begin with brief definitions of both the key terms, and perhaps say no more than it is widely assumed that both development and urbanisation go hand in glove. As countries develop, so they become more urbanised. Why should this be?

One good framework to follow would be to take the five strands of urbanisation (Figure 9) and show how each relates to development:

- Economy: shift to secondary and tertiary activities, and rising prosperity. Development raises food supply (advances in agriculture; importing food becomes possible). People free to work in other sectors. Rising prosperity creates expanding market for goods and services. Mexico (*Case study 6*) shows this sectoral shift, encouraged by availability of mineral and energy resources. Role of government and release of rural population well illustrated by China (*Case study 5*).
- Population: rural–urban migration. Encouraged by the release of people from food production and the concentration of manufacturing and services in accessible locations. Well shown by Ethiopia (*Case study 4*), but with other factors, not connected with development, helping to increase movement to towns and cities.
- Settlements: the growth and concentration of secondary and tertiary activities, plus rural–urban migration, creates new settlements (towns and cities) and raises the scale of those settlements (emergence of primate cities, e.g. in Ethiopia, Mexico, Iraq).
- Environment: prevalence of bricks and mortar; more pollution. The inherent nature of settlements produced by the sectoral shift and rural–urban migration. Reference to problems of Mexico City (*Case study 13*), Addis Ababa and Dhaka (*Case study 26*).
- Lifestyle: change in occupation and environment responsible for shifts in values and behaviour. Again, Addis Ababa, Mexico City and Dhaka could be used to provide support.

A possible conclusion to be drawn from this is that the main link between urbanisation and development is the economic one:

- increased agricultural productivity and the release of workers to take up jobs in other sectors
- the rising demand for goods and services

Why not try writing the essay now, using the above framework.

The urbanisation pathway: later phases

Agglomeration and suburbanisation

There is a critical stage along the urbanisation pathway when the processes of centralisation (agglomeration) begin to weaken. They are gradually superseded by other processes, some of which operate in the opposite direction (Figure 15). **Centrifugal** forces may already be at work, even in the early phases of urbanisation. For example, while the agglomeration that leads to the emergence of primate mega-cities like Addis Ababa and Mexico City is inherently centripetal, the outcome is the gradual extension of the built-up area. It is easy to see the link between this accretion and the first main process of decentralisation: **suburbanisation**. This is simply the outward spread of the built-up area to engulf surrounding rural areas and settlements. The process is fuelled by:

- rural–urban migration
- natural increase
- setting up new businesses and services
- decentralising people, employment and services from the central and inner areas of the town or city

Counterurbanisation

Counterurbanisation is a widely misunderstood term. It may be that the fault lies in the word itself, which implies change that is 'anti-urban'. This is only partly the case. Although one of the components of counterurbanisation is urban–rural migration (Figure 15), only a proportion of these migrants are turning their backs on urbanism. Many retain urban-based jobs by commuting, while others remain patrons of urban-based services. As a result, there is rural dilution.

A more significant component of counterurbanisation is made up of people and businesses opting to move down the urban hierarchy to smaller towns and cities. This is not 'anti-urban' — it is merely 'anti-big city'. This form of counterurbanisation leads to further urbanisation by diffusing growth down the urban hierarchy.

Figure 15 *The processes of urbanisation and their spatial impact*

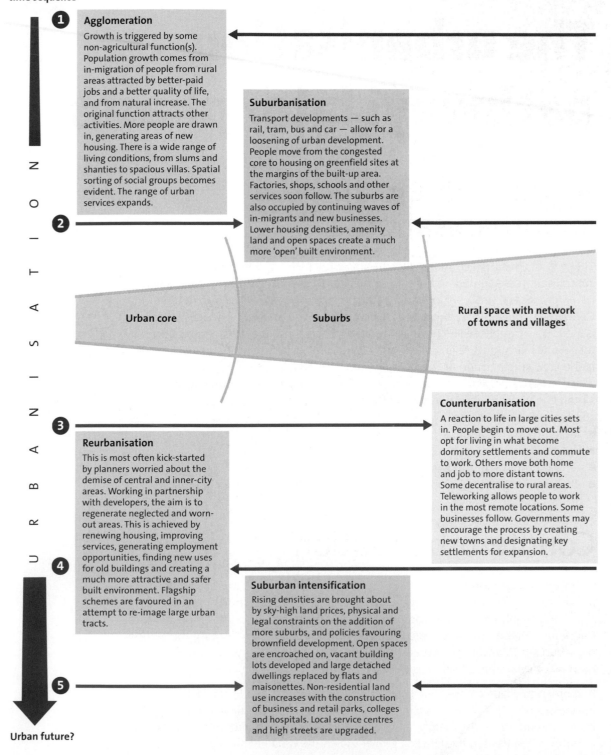

Process:
time sequence

Centralising (centripetal)
Decentralising (centrifugal)

U R B A N I S A T I O N

① Agglomeration

Growth is triggered by some non-agricultural function(s). Population growth comes from in-migration of people from rural areas attracted by better-paid jobs and a better quality of life, and from natural increase. The original function attracts other activities. More people are drawn in, generating areas of new housing. There is a wide range of living conditions, from slums and shanties to spacious villas. Spatial sorting of social groups becomes evident. The range of urban services expands.

Suburbanisation

Transport developments — such as rail, tram, bus and car — allow for a loosening of urban development. People move from the congested core to housing on greenfield sites at the margins of the built-up area. Factories, shops, schools and other services soon follow. The suburbs are also occupied by continuing waves of in-migrants and new businesses. Lower housing densities, amenity land and open spaces create a much more 'open' built environment.

Urban core   **Suburbs**   **Rural space with network of towns and villages**

Counterurbanisation

A reaction to life in large cities sets in. People begin to move out. Most opt for living in what become dormitory settlements and commute to work. Others move both home and job to more distant towns. Some decentralise to rural areas. Teleworking allows people to work in the most remote locations. Some businesses follow. Governments may encourage the process by creating new towns and designating key settlements for expansion.

Reurbanisation

This is most often kick-started by planners worried about the demise of central and inner-city areas. Working in partnership with developers, the aim is to regenerate neglected and worn-out areas. This is achieved by renewing housing, improving services, generating employment opportunities, finding new uses for old buildings and creating a much more attractive and safer built environment. Flagship schemes are favoured in an attempt to re-image large urban tracts.

Suburban intensification

Rising densities are brought about by sky-high land prices, physical and legal constraints on the addition of more suburbs, and policies favouring brownfield development. Open spaces are encroached on, vacant building lots developed and large detached dwellings replaced by flats and maisonettes. Non-residential land use increases with the construction of business and retail parks, colleges and hospitals. Local service centres and high streets are upgraded.

Urban future?

While it is important to recognise these two components of counterurbanisation, we need to be aware of the two-fold distinction that exists between spontaneous (voluntary) and planned counterurbanisation. Both socialist and capitalist countries have tried to encourage decentralisation from large, overgrown cities. Examples are the British New Towns and regional development programmes which, after the Second World War, attempted to:

- 'slim down' the conurbations — particularly Greater London
- reduce the magnetism of London and the southeast to people and businesses
- redress some of the imbalance between the regions

Reurbanisation and suburban intensification

Two more urbanisation processes need to be identified. Both involve an element of 'new interest' in parts of the built-up area, and both are encouraged where:

- the built-up area is constrained by some form of greenbelt or break
- policy decrees that **brownfield** sites should be developed in preference to **greenfield** sites

The first process is known as **reurbanisation**. It focuses on reusing abandoned or neglected areas in the central and inner city (Figure 15). It is a centralising process involving the following types of activity:

- gentrification — the inward movement of high-income and middle-class people
- finding new uses for buildings and spaces
- changing building densities
- encouraging particular social or ethnic groups
- renovation — giving tired urban landscapes a make-over
- **re-imaging** large urban tracts

Early examples of reurbanisation included the postwar **renewal** of CBDs and the **redevelopment** of inner-city areas of slum housing.

The second process has also been in operation for some time, but its collective impacts are only just beginning to be recognised. There is still no official term or name for the process, but we have called it **suburban intensification** (Figure 15). It has two distinctive features:

- It affects the **suburbs**.
- It involves an intensification of land use.

Its diagnostic features include:

- raising housing densities by infilling vacant building plots, developing public open spaces (e.g. parks and playing fields), and replacing large, detached houses with flats and maisonettes
- squeezing in business and retail parks, hospitals and colleges, which often involves taking over amenity land
- upgrading local shopping areas by allowing 'high-street' names such as Tesco, Boots and the Carphone Warehouse to replace small family businesses

The outcome of these changes is that the suburbs begin to lose their 'openness' and become more urban.

Figure 15 raises three further points:

- These five processes — agglomeration, suburbanisation, counterurbanisation, reurbanisation and suburban intensification — are shown in a time sequence. There would be widespread agreement about the start times of the first two processes, but what about the last three? It is likely that many would agree with the order shown. However, the reality may be that they started more or less at the same time. In the UK, this may have been as early as the 1960s (*Case study 11*).
- It would be wrong to think that a new urbanisation process begins only when its predecessor has finished. We need to recognise that several of these processes may be at work in a town or city at a particular time. The important aspect is the relative strengths of those processes and their impacts. At a national level, those relative strengths determine whether the national urbanisation curve continues to rise, levels off or begins to dip.
- Counterurbanisation, reurbanisation and suburban intensification are all **post-industrial** developments. Their rise is linked to **deindustrialisation** — the broad shift from manufacturing to services.

Case studies 8–11 explore the later stages and processes of the urbanisation pathway. Each represents a different political context:

- where strong government leadership has created a formidable city state (Singapore — *Case study 8*)
- where a transition is being made from socialist government control to market forces (Russia — *Case study 9*)
- where market forces reign supreme (the USA — *Case study 10*)
- where government direction co-exists with market forces (the UK — *Case study 11*)

Case study 8 **SINGAPORE**

A city state

Singapore is unusual in a number of respects. In development terms, it was until recently rated as a newly industrialised country (NIC), but its level of urbanisation was already 100%. Therefore, it appears that the general correlation between development and urbanisation does not apply here. The explanation for the discrepancy probably lies in Singapore's small size (it is smaller than New York City). It occupies a modest island and many small ones, just off the southern tip of the Malaysian peninsula (Figure 16). Its total land area is just less than $650\,km^2$ and its present population is just over 4 million. It is the most densely populated nation in Asia (6485 people per km^2). Furthermore, its per capita gross national income (GNI) is roughly the same as the USA — it is richer than most European nations. Singapore is truly a city state.

Singapore started life as a British colonial trading post. Since independence in 1965, it has been transformed radically from a low-rise colonial port to a high-rise post-industrial city state. In the process, most of the countryside has been converted into urban space. This shortage of space has been so severe that land has been reclaimed from the sea (see p. 41). The rise of Singapore as the most prosperous and successful of the 'Asian Tigers' is truly remarkable. Not only is it incredibly small, but it has to import all of its water, most of its food, and all of its energy and industrial raw materials.

In response to clear government direction, the economic development of Singapore has gone through four stages:

- Up until the mid-1970s, export-oriented industry was encouraged, as well as foreign investment by TNCs. This strategy was highly successful.
- Industrial restructuring began in the late 1970s, which involved a shift from labour- to capital-intensive industries. The result was much automation and considerable broadening of the product base.
- From the mid-1980s to 2000, there was a sector shift from manufacturing to services.
- Since 2001, as a result of stiffening competition from lower-cost countries for goods and services, the government has promoted a knowledge-based economy driven by ICT and innovation. Sectors of the Singapore economy that were closed previously to foreign investment, such as telecommunications and financial services, have now been opened up.

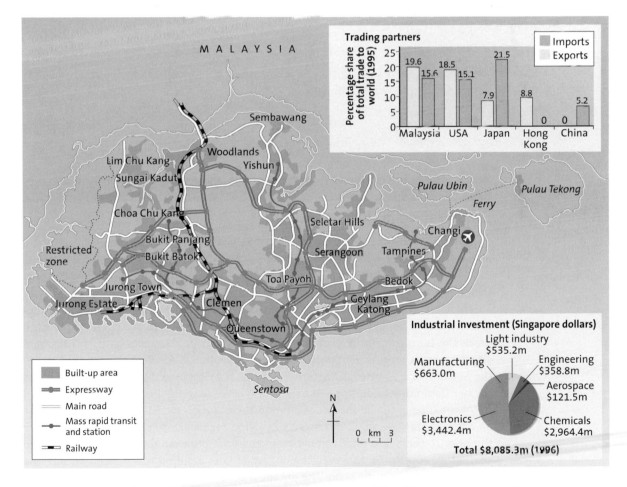

Figure 16
Singapore: urban and economic development

Singapore's incredible success as a modern city state can be explained by:

- **its natural harbour and related port functions**. Singapore is the most dependent of all the Asian Tigers on overseas trade. It is the world's busiest port, handling one-third of the world's trade, with one ship docking every 2 minutes.
- **the quality of its human resources**, who are well educated, hardworking, enterprising and ambitious. However, there is a shortage of unskilled labour and professional people.

■ **the firm leadership provided by the Singapore government**, although this includes a tight control on behaviour, restricted freedom of the press and a largely one-party democracy.

In 1965, Singapore was suffering severe residential overcrowding, combined with problems of unemployment and slum living. Massive **social housing** and industrialisation programmes were launched. These involved clearing slums, rebuilding obsolete buildings and rapid development of much of the green space remaining on the island. Singapore was soon transformed into a densely built modern city, with a characteristic skyscraper and concrete landscape. However, a recent programme to create and maintain green spaces is gradually turning this concrete city state into a garden city state.

8 Using case studies

Question
Singapore is an unusually small country. What do you think is unusual about it in terms of urbanisation?

Guidance
There are at least four features you should be able to identify. Singapore is unusual in terms of urbanisation in the following respects:

■ the apparent discrepancy between the levels of urbanisation and development

■ the strong government direction of a market economy
■ the significance of location as a resource
■ the quality and efficiency of the built environment

Case study 9 — RUSSIA

Urbanisation in fits and starts

In development terms, the exact status of Russia is not clear. As the most powerful and influential member of the former Union of Soviet Socialist Republics (USSR), Russia would claim to have been an MEDC. It is an oil-rich state and this bodes well for its future. However, since 1991 when the USSR broke up and Russia became an independent state, it has slipped back in terms of development indicators such as per capita gross national income (GNI). This slippage has occurred as Russia struggles to make the transition from a command economy to a market economy. This unusual situation explains why the World Bank recognises the 15 former Soviet republics and the Soviet bloc countries of Eastern Europe as making up a distinct group known as Former Communist Countries (FCCs).

At the time of the Bolshevik or Communist Revolution in 1917, European Russia was an agrarian state — only 15% of its 13.2 million people lived in towns and cities. Over the next 70 years, the total population increased more than ten times and the urban percentage rose to 74. However, the history of Russia during this period was not one of uninterrupted urbanisation. The decade after 1917 was marked by a depopulation of the cities, and the effect of the Second World War was the same.

There was intensive urbanisation between 1930 and 1940, as well as between 1950 and 1965. In 1926, there were only 26 million urban dwellers, but by 1990 the number was 109 million. These two surges in urbanisation coincided with the forced transfer of rural people to towns and cities. The collectivisation of farming allowed the government to lay its hands on surplus peasant farm labour and move it to industrial cities.

After 1965 and until the collapse of the Soviet Union, there was a steady rise in the level of urbanisation. The main cause was the government's direction of industry — in particular, those involved directly and indirectly in making armaments. This was at the height of the Cold War between the USSR and the West. The USSR was also trying to extend the communist world by force, particularly in south and southeast Asia (e.g. Vietnam). This was helped by the significant surge in population growth as the country recovered from losing millions of its men during the Second World War.

Since 1991, the level of urbanisation has flattened out at 73%. This is the outcome of several factors:

- a profound downturn in the economy
- the onset, for a while, of reverse migration — urban-to-rural
- a low birth rate

St Petersburg

The building of St Petersburg started in 1703, when Tsar Peter I decided that this part of Russia needed an outlet to the sea and a port for trade via the Baltic. The site chosen was the delta of the Neva River, with its many islands and distributaries. This patchwork of islands, channels and bridges means that St Petersburg is known as 'the city on 101 islands' or 'the Venice of the North' (Figure 17).

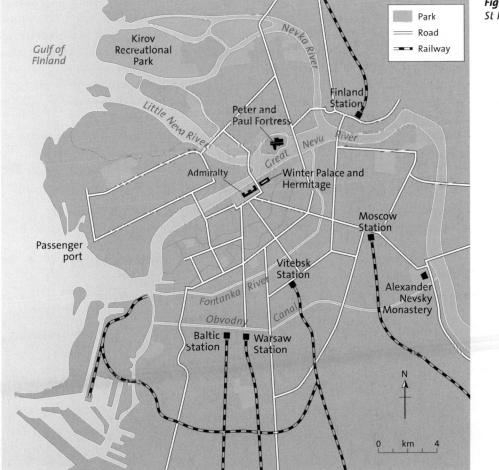

Figure 17
St Petersburg

Italian and French architects designed the city, giving it the spacious, classical beauty it has retained. In 1712, the new city was given a boost when it became the capital. The Tsar required the aristocracy to move there, to build lavish homes for themselves and to contribute to the cost of constructing the government buildings.

St Petersburg soon replaced Archangel as Russia's leading sea port. It also became an important commercial centre. From the second half of the eighteenth century, it was the main industrial centre — at first for shipbuilding and engineering, and later for textiles. It gradually became one of the world's most vibrant capitals and cultural centres, renowned as a scene of lavish and reckless social life.

However, the seeds of social unrest had been sown. In 1917, the Bolshevik Revolution launched nearly 80 years of communist rule. In the following year, the capital function was returned to Moscow. In 1924, in an attempt to remove the city's close ties with the Russian royal family, it was renamed Leningrad.

During the Second World War, the city was cut off from the rest of the Soviet Union. It was besieged by German forces for over 2 years, during which time many hundreds of thousands died of famine and disease. Much damage was also inflicted on the city.

The Socialist years have left their mark on the urban landscape, with an encircling ring of densely packed, poorly built, high-rise apartment blocks. Beyond this ring are industrial zones. Fortunately, much of the war-time damage to the architectural **heritage** of the eighteenth and nineteenth centuries was repaired. Since the collapse of the communist regime, many fine historic buildings have been renovated. Renovation is now taking place in the areas of upper- and middle-class housing in the centre, as well as in the surrounding belt of working-class housing dating from the pre-Socialist period. At the margins of the city, there are also signs of change, with new housing developments (mainly for owner-occupiers). They have a lower profile than the high-rise apartment blocks of the Socialist period, and the quality of their construction is significantly better.

Having reverted to its original name, the city now attracts a growing volume of Russian and foreign tourists. They are drawn by buildings such as the Admiralty and the Winter Palace (Figure 17). Tourism, together with industrial, transport, scientific and cultural functions, helps to support a city of 5 million people.

This case study illustrates the following points:
- *As with Iraq, a country can slip back along the development and urbanisation pathways.*
- *The impacts of war and changing political systems leave their mark.*
- *Cities come into being to meet particular needs.*
- *The subsequent economic value of a city's heritage can be realised.*

9 Question

Read *Case study 9* again, particularly the last paragraph. Use the information to draw a simple diagram to show the broad structure of the city. Include some annotations about the urban processes at work.

Guidance

Copy Figure 17 and use it as your base map. It may help you to refer to Figure 35 on p. 59.

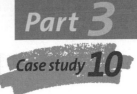

THE USA

Nearing the peak?

The urbanisation of the USA has taken place over the last 200 years. Urbanisation spread from east to west as part of the broad settlement and colonisation of the country largely by immigrants from Europe. Five major stages can be recognised in the evolution of this urban system (Figure 18):

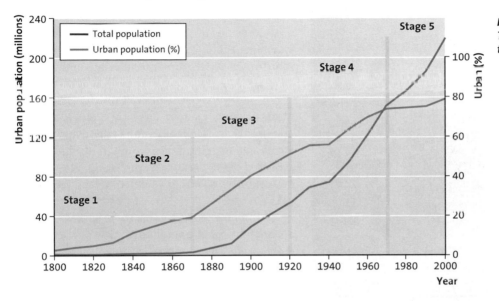

Figure 18
The urbanisation of the USA, 1800–2000

- **Stage 1 (pre-1830):** the location and early spread to towns and cities was influenced by transport modes of the time, such as horse-drawn wagons and sailing vessels.
- **Stage 2 (1830–70):** the development of railways, canals and steamships helped the spread of urbanisation. Despite the debilitating effects of the Civil War (1861–65), the percentage urban figure rose from 7 to 20.
- **Stage 3 (1870–1920):** there was a marked take-off in urbanisation due to large-scale industrialisation, massive immigration and further significant improvements in transport. This is when the basic urban system was established. For the first time in its history, most Americans were classified as urban.
- **Stage 4 (1920–70):** during this period, motor cars reshaped US towns and cities, mainly through their encouragement of suburbanisation. Air travel reinforced the already established urban pattern. The Great Depression of the 1930s caused the rate of urbanisation to ease.
- **Stage 5 (1970–today):** suburbanisation continued to increase the national built-up area, but counterurbanisation also gathered momentum. It began to look as if the urbanisation curve was levelling off, but there was a small recovery in the 1990s. The country's urban centre of gravity continues to move decisively to the west coast.

Los Angeles

Los Angeles (LA) has a population of around 14.5 million people, making it the second largest urban agglomeration in the USA. Compared with New York, LA is not only younger (its origins date back only to 1781), but its form of growth is different —

Figure 19
*Los Angeles —
a product of the
motor car*

horizontal and sprawling. This difference is down to the fact that New York was created by the railway and steamship, whereas LA was the product of the motor car (Figure 19). Also encouraging the low-rise character of the city is the ever-present earthquake risk.

After LA was linked to the rest of the country by rail (1885) and an artificial port built (1914), the city began its meteoric rise. This rise was fuelled by a succession of lucrative economic activities spanning the film industry and entertainment, tourism, oil, modern manufacturing, and cutting-edge aerospace and other high-tech industries.

LA today is a cosmopolitan city. People of Mexican extraction form LA's largest ethnic minority and high birth rates mean the Latino population is increasing in importance. Immigrants have been attracted from all parts of the world. People from 140 countries, speaking 96 different languages, now call LA home.

LA is renowned as a gargantuan sprawl, covering an area of 88 000 km^2. **Edge city** developments (*Case study 14*) continue to enlarge the built-up area. There was concern that this persistent decentralisation would drain the downtown area of both people and business, but this has not occurred. Instead, **downtown** LA is fast becoming a cultural and entertainment centre, with such flagship developments as Staples Center (a basketball arena), the Walt Disney Concert Hall, the California Science Center and the new cathedral. Success in cultural tourism is beginning to have a ripple effect, with a revival of the property market in adjacent inner-city areas.

Chattanooga

Chattanooga is a modest city of less than half a million, tucked away in the southeast corner of the state of Tennessee, close to the border with Georgia. Its strategic position as a railway junction made it an important Northern (Union) stronghold during the Civil War. It grew into a major rail and road transportation node and a centre for manufacturing cotton textiles and metals, but today, the city has ceased to be an industrial town. In 1969, it earned the unenviable reputation as the 'dirtiest city in the USA'. However, this has now changed and the city relies on tourism, which is flourishing thanks to its:

■ good accessibility via interstate highways
■ associations with the civil war
■ architectural heritage
■ immediate hinterland, with fine scenery and outdoor recreational opportunities

What is so special about this modest city and why is it so well known outside the USA? The answers can be found in *Case study 32*.

The main point raised by this case study is that the USA is the supreme example of a free-market economy. Planners in general have been unable to exercise the same degree of control over urbanisation and its impacts as in some other MEDCs. This may explain why the processes of reurbanisation and suburban intensification are barely discernible. After all, the one thing that the USA has plenty of is space. There has not been the same need to protect greenfield areas and target brownfield sites as in the UK (Case study 11). The USA is recognised universally as leading global development, so why is it not the most urbanised?

THE UK

Over the top?

As the birthplace of the Industrial Revolution, Britain can rightly claim to have been the world leader along the urban pathway. Industrialisation provided the basic fuel for the take-off in mass urbanisation. At the time of the first census in 1801, around 50 years after the start of the Industrial Revolution, Britain was already a highly urbanised country. For much of the last 200 years, suburbanisation has been a dominant process, adding thousands of square kilometres of low-density, built-up areas.

Figure 20 shows that England may have peaked in terms of the level of urbanisation. Counterurbanisation is a recognisable process, with decentralisation taking place from larger cities. People and businesses are relocating to lower-order urban settlements and rural areas.

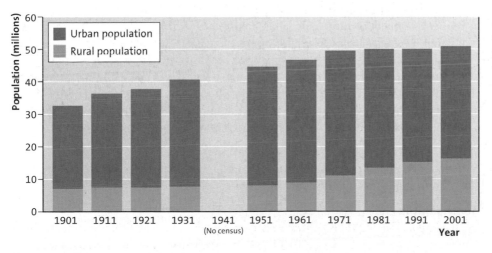

Figure 20 *England and Wales: a century of urbanisation, 1901–2001*

London

The ability to cross the River Thames at a point accessible to sea-going ships strongly influenced the location of London in Roman times. For most of the next 2000 years, it served as the national capital, which helped to boost its primacy. In 1801, London was eight times larger than its nearest rival, Manchester. Today, that primacy has been reduced to a factor of three.

London was the world's first millionaire city, passing the threshold in 1810. Its status as a global city was established during the nineteenth century, when Britain was the leading industrial nation. The British Empire was at its peak and extended over much of the world. London became a vital hub in the emerging global economy of the day.

Although the British Empire has disappeared and Britain is no longer the industrial nation it once was, the momentum has been such that London is still recognised as one of the leading financial, legal and cultural capitals of the world. However, it is no longer one of the world's largest urban agglomerations — the population of its continuous built-up area (the Greater London conurbation) levelled off after 1951, at around 8 million (so technically it is no longer a mega-city).

The following processes have been involved in the evolution of London's built-up area since the mid-nineteenth century:

- **Suburbanisation** has added areas of new housing, industrial estates, business parks and retail parks inside the green belt. This outward sprawl of the built-up area was made possible by developments in transport, with improvements in public transport services by rail and bus, as well as by increased car ownership. These advances allowed increasing distances to be travelled between home and place of work, and the age of commuting began.
- **Counterurbanisation** is the informal and voluntary movement of people and businesses to smaller towns and cities. This has taken place mainly within the southeast, and has reached further afield, even to rural areas.
- **Planned decentralisation** of people and jobs to new and expanded towns has been part of the same process. Again, the dispersal has mainly been in the southeast.
- **Reurbanisation:** since the 1970s, planners have persuaded developers to reuse brownfield sites, particularly in the inner city. Flagship redevelopments, such as Canary Wharf, are the outcome of this process.
- **Suburban intensification:** given the combination of a restraining green belt and a government policy favouring brownfield over greenfield sites for new urban development, many of London's suburban areas have been experiencing this important change. Residential densities are being increased by building flats rather than houses. One- and two-bedroom flats now account for 80% of all new homes built in Greater London. This shift is raising fears that we are creating new urban ghettos of tiny flats suitable only for the young. The attractions of the shift are higher profits for the developers and meeting another government policy to provide **affordable homes**. However, there is now a growing shortage of family homes.

While all the above processes continue today, the balance has shifted away from suburbanisation towards the other processes.

Southampton

With a population of 221000, Southampton is best described as a middle-order regional city. Its roots go back to Saxon times, when it first developed as a port. During the Middle Ages, it prospered on trade along the coast and across the English Channel. Its status as a port was given a great boost in the early nineteenth century with the completion of a railway link with London (Figure 21). This allowed for faster travel between the two cities than by sea. Southampton soon became the country's leading passenger port, particularly for transatlantic services and for destinations to those parts of Africa, Asia and Australasia that were part of the British Empire.

Southampton continues to flourish as a port, but the nature of its traffic has changed. It is now a leading container port, i.e. its trade is dominated by cargo. The city is also the centre of the UK's cruise industry. The people who use the port are tourists rather than passengers on scheduled services.

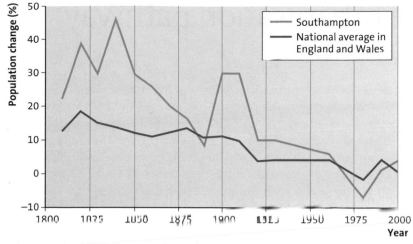

Figure 21 Rates of population change in Southampton, 1800–2000

As the principal city in central southern England, Southampton is an important 'central' place, providing a good range of tertiary (retailing, media and tourism) and quaternary (business, finance and education) services for a large part of southern England — a catchment area of 4 million people. Of the 110 000 people working in the city, 85% are in services and only 10% are in manufacturing. The shift to the service sector is part of a re-imaging of the city, led by the city council and the South East England Development Agency (SEEDA).

Like other British cities, Southampton experienced suburbanisation during the twentieth century. However, the city is now physically confined by a tightly drawn green belt. With the city economy continuing to expand, the inevitable has happened. New, detached suburbs have been built on the other side of the strategic gap, such as at Eastleigh, Chandlers Ford and in the Waterside Parishes. There are parallels with what has happened in and around London. Those similarities include the onset of reurbanisation and suburban intensification (*Case study 19*).

The points raised by this case study are as follows:
■ *The UK has probably peaked in percentage urban terms. This suggests that counter-urbanisation to rural areas is beginning to make its mark.*
■ *This peak has been encouraged by government policies that actively discourage the built-up area from spilling over into the countryside.*
■ *Both Southampton and London, although differing enormously in size and reputation, continue to experience the same processes typical of the later phases of the urban-isation pathway.*

10 **Using case studies**

Question

How do the two global cities of Los Angeles and London compare? What are their similarities and differences?

Guidance

Assess both cities against the attributes shown in Figure 8, p. 7.
Websites: **www.ci.la.ca.us/**
http://en.wikipedia.org/wiki/Los_Angeles
www.london.gov.uk

The urbanisation pathway as a whole

That completes coverage of the later phases of the urbanisation pathway and the associated processes (Figure 22). The main features of the pathway as a whole are as follows:

■ It is a model or generalisation — obviously there will be some exceptions.
■ The pathway is a broad one — not all countries or cities will follow in exactly the same footsteps. There will be subtle divergences as countries weave their way in the same general direction towards a better quality of life for more people.
■ It is about shifts in the relative importance of processes that are, for most cases, happening at the same time.
■ Countries and cities move along the pathway at different paces and at speeds that do not have to be consistent over time. The pace of development is probably the key factor. In some cases, the role of government, both national and local, is becoming increasingly significant.
■ Progress along the pathway can also vary between different regions and cities within a country.
■ Movement along the pathway is not always forwards — countries and cities can slip backwards too.
■ Just as it is unlikely that all countries will one day become MEDCs, so is it unlikely that all countries will reach the end of the pathway as we see it today.
■ The pathway will continue to extend into the future, so new processes and characteristics will emerge.

Figure 22 shows where the countries and main cities chosen as case studies in this book are currently located along the urban pathway.

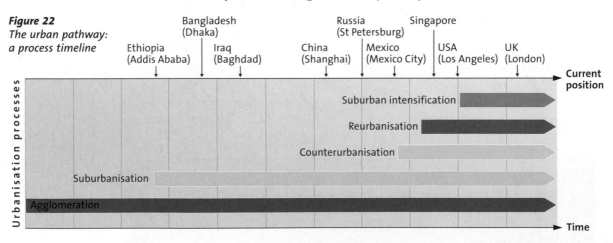

Figure 22
The urban pathway: a process timeline

The outcomes of the different processes along the pathway can vary considerably across the world. What the urban environment or landscape looks like, and what the issues are that confront their managers, are major topics explored in parts 4 and 5.

ging cities

The urbanisation pathway in parts 2 and 3 is a generalisation, based on broad similarities shared by countries as they develop and become more urbanised. As they move along the pathway, most countries appear to undergo the same sequence of process change (Figure 22, p. 36). The following questions will be explored in this chapter:

- At any point along the pathway, will those processes always produce the same sort of urban environment?
- Will the same spatial patterns of land use and people be found within the built-up area, no matter where you are in the world?
- Will the challenging issues also be the same?

The spatial patterns of cities are mosaics, made up of many small areas. Each area has an individual character. It is distinctive because of a particular combination of characteristics, such as building type, age, density of development, activities or types of people living there. The boundaries between these areas are not always clearly defined — there are transitions from one area to the next. Urban spatial patterns are explored in many human geography textbooks. However, these so-called models of urban structure are frequently misunderstood, particularly in terms of their purpose.

Given the great diversity of cities around the world, it would be wrong to think that one generalisation or model could be devised to cover them all. The best-known urban models — concentric zone, sector and multiple nuclei, together with Mann's fusion of the first two — were intended only as generalisations regarding MEDC cities. Furthermore, they were first proposed decades ago, so they are now out of date. These models have little to do with **post-industrial cities**. Think about the development and urbanisation pathways and the way in which countries are placed along them (Figure 22). The mix of urbanisation processes at different points along those pathways may well produce different types of urban pattern. Consider also the likely differences in the spatial patterns of those cities governed by socialist principles and those driven by market forces. Might culture also be a differentiating factor? For example, cities that have flourished in Islamic culture are different from Christian or Western cities.

One of the main purposes of a model is to put the spotlight on factors and processes that seem to be major influences in the shaping of city patterns (Figure 23). Is the main factor:

- distance from the city centre — emphasised by the concentric zone n...
- the attraction of main routes — emphasised by the sector model?
- the power of agglomeration — emphasised by the multiple nuclei model?
- decentralisation of functions and people?

Are other factors, such as size, site conditions or inertia, at work? How influential is government direction at national and local levels?

Urban models have another limitation, in addition to their restricted validity, in terms of space and time. They tend to suggest that the city pattern is static; but cities are highly dynamic. They are constantly changing in response to external and internal processes. External processes are mainly those discussed in parts 2 and 3, with the notable exceptions of post-industrialism and globalisation. Do internal processes of change have a greater impact on city patterns?

We will focus on three forces affecting the city pattern (Figure 23):
- the physical environment
- internal processes of change
- economic globalisation

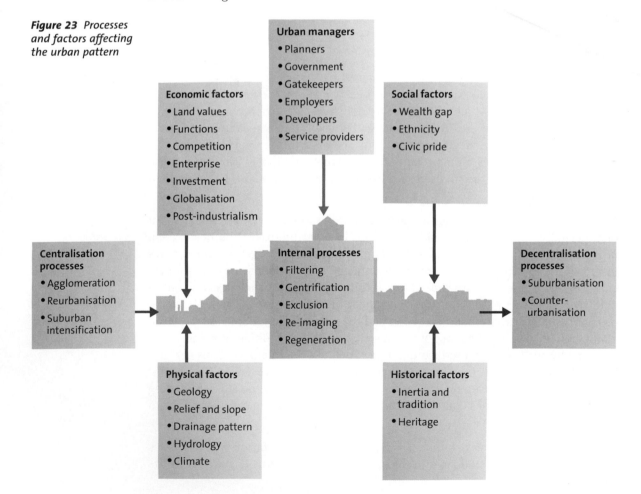

Figure 23 *Processes and factors affecting the urban pattern*

Urban managers
- Planners
- Government
- Gatekeepers
- Employers
- Developers
- Service providers

Economic factors
- Land values
- Functions
- Competition
- Enterprise
- Investment
- Globalisation
- Post-industrialism

Social factors
- Wealth gap
- Ethnicity
- Civic pride

Centralisation processes
- Agglomeration
- Reurbanisation
- Suburban intensification

Internal processes
- Filtering
- Gentrification
- Exclusion
- Re-imaging
- Regeneration

Decentralisation processes
- Suburbanisation
- Counter-urbanisation

Physical factors
- Geology
- Relief and slope
- Drainage pattern
- Hydrology
- Climate

Historical factors
- Inertia and tradition
- Heritage

The nature and impact of the physical environment

The relationship between the physical environment (site conditions) and the built-up area is reciprocal. The impact of the physical environment may be most important during the early formative years. The built-up area is strongly moulded, in a negative way, by site features such as waterfronts, steep slopes and marshy ground.

Because of better technologies and the ever-rising price of urban land, the grip of the physical environment has weakened, so that urban growth seems possible almost anywhere. Today, the physical environment is often exploited in the re-imaging of cities, for example, in St Petersburg (*Case study 12*) and Shanghai (*Case study 29*). However, global warming now looms, challenging the view that the human hand is uppermost. Rising sea-level poses a real threat to the prosperity — and maybe even the survival of — many of the world's cities. A significant number of leading cities have a coastal location (Figure 6, p. 5).

Case study 12 looks at how cities have won land from the sea to build waterfronts. How long will it be before the sea claims back that land?

THE WATERFRONT

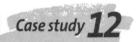

Case study 12

An urban frontier and zone in transition

Although the water's edge is a frontier — a feature constraining urban growth — it can also be an important stimulus. From the earliest times, the port and ferry potential of waterfront locations has been exploited by towns and cities. In most cases, the original waterfront has been gradually, but substantially, modified. This has been for reasons ranging from flood control and storm protection, to better access for shipping and the reclamation of land for port and other uses.

St Petersburg

This planned city has had to cope with a highly fragmented site, made up of many small islands (*Case study 9*). This serious constraint had to be overcome if St Petersburg was to emerge as Russia's Baltic port. The problems were gradually overcome by:

- joining small islands together by reclaiming the narrow channels that separated them
- building many bridges to link large islands

Ironically, the island character of the city is now an asset rather than a drawback (Figure 17). The city's reputation as 'the Venice of the North' is now attracting a growing number of tourists.

Southampton

Southampton was originally located on the narrow peninsula between the rivers Itchen and Test, as they converge to form Southampton Water (*Case study 11*). Throughout its subsequent history, this location has presented the city with two challenges: accessibility (crossing the two rivers) and space for expansion (particularly of its port installations). For centuries, the peninsula was artificially extended by reclamation to create new docks

(*Case study 19*). Much of this land is now brownfield sites, with residential, leisure and retail developments replacing old warehouses and redundant docks.

Figure 24 shows the extent of mudflat reclamation along Southampton Water. The process continues to this day, to create much-needed space for the port's flourishing container business. It has ignited the recent controversy over the proposed expansion of the container terminal onto the other side of the River Test at Dibden Bay.

Figure 24
Wetland reclamation in and around Southampton

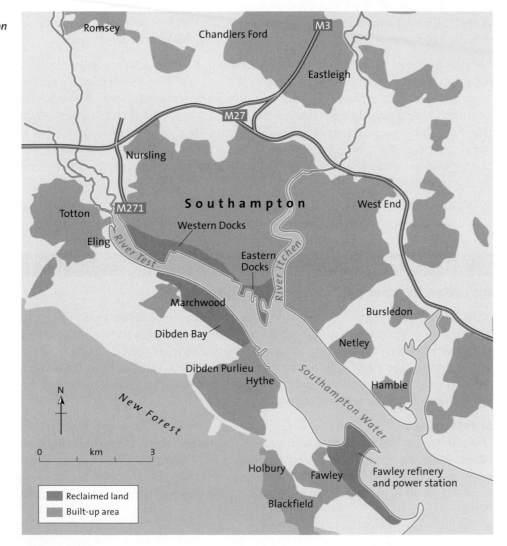

Singapore

The area of mudflat and wetland reclaimed around the shores of Singapore Island is considerable (Figure 25). The coastline was once broken by a number of estuaries, but these have been reduced and converted into land. Early land reclamation focused on the vicinity of the main urban area. This was primarily aimed at pushing the port's shoreline towards deeper water and providing new docks. However, some draining and infilling of swampy areas took place for health reasons, and to provide space for more buildings.

Since independence in 1965, economic factors and population pressure have brought about a shift to large-scale reclamation projects. Over 50 km^2 of 'new' land has

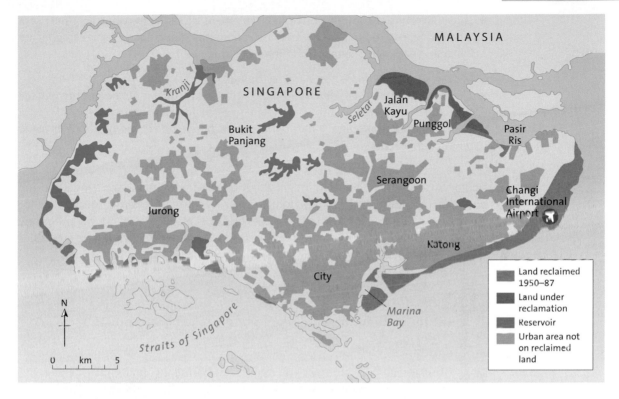

Legend:

- Land reclaimed 1950–87
- Land under reclamation
- Reservoir
- Urban area not on reclaimed land

been won from the sea. The outline of Singapore Island today differs greatly from that of 50 years ago. These changes are ongoing, as the prosperous city state continues its insatiable appetite for still more space.

Figure 25
Singapore: urban pressures and land reclamation

The waterfront today

For centuries, waterfronts have been crucial to the prosperity and growth of many major cities, mainly because of their importance to the port function and port-related industries, such as oil-refining, milling and steel. However, in the post-modern (post-industrial) era, it is being realised that waterfronts can provide new and exciting opportunities. These opportunities arise because of the:

- relocation of modern port activities, especially container handling and storage, to more spacious locations away from the early docks
- decline in port-related industries
- removal of those remaining industries, especially the polluters, to locations away from residential areas

These changes have released waterfront space for new ventures. These enterprises can be placed into four categories:

- Those involving a major shift in land use, such as at Canary Wharf in the London Docklands and Darling Harbour in Sydney, Australia.
- Those acting as the catalysts for a wider urban **regeneration**. The redevelopment of the vacated waterfront has a ripple effect that spreads to adjacent parts of the city. This scenario is well illustrated by Baltimore in the USA.
- Those providing the opportunity to 're-expose' the historic nucleus and exploit its heritage and tourism potential. This is shown by Havana in Cuba and Amsterdam in the Netherlands.

- Those involving a comprehensive redevelopment of the waterfront as a part of a much wider city re-imaging exercise. Two outstanding examples in this category are Bilbao in Spain and Shanghai in China (*Case study 29*).

This case study illustrates the following points:
- *The pressure for urban land and the need for land-water interchange can change the natural environment.*
- *The waterfront is essentially part of the **urban fringe** and in many cities it is being revalued and redeveloped.*

The devastation of New Orleans by Hurricane Katrina in 2005 (*Case study 28*) is a salutary warning to those who might think that today's cities have conquered the physical environment and natural forces. The risk element remains, in this case for a city that is not only located below sea level, but is also close to a hurricane-prone coast and on the banks of a major river liable to flood. New Orleans was a disaster waiting to happen.

11 Question
List the costs and benefits of using and extending waterfronts.

Guidance
Use *Case Study 12* but refer also to *Case Studies 19*, *28* and *29*. Work under the headings 'environmental', 'economic' and 'social'.

Website: **www.planningsummerschool.org/papers/year2003/2003B017AU.pdf**

The nature and impact of internal processes

The character of individual cities is not only affected by global location and site conditions. Processes of change also make their mark (Figure 23). These processes are largely of a physical, economic or social nature. They fall into two broad categories:
- Processes that are **external** to the city, of which economic development is by far the most important. These provide the basis for urban growth.
- Processes that are **internal** within the city, which are largely related to the particular ways in which cities grow outwards and how structures and patterns change. There are three sets of internal processes:
  – suburbanisation
  – filtering, gentrification and exclusion
  – regeneration and re-imaging

These processes are important influences on both the general form of the city and its social geography. Additional processes at work within the city include those relating to the spatial sorting of economic activities and different land uses within the built-up area. The urban land market and planning are two powerful influences on the city's economic map (Figure 23, p. 38).

The most significant process of urban growth is suburbanisation. This results in rings of new growth around the edge of the built-up area, as shown in the concentric zone model. Suburbanisation on a large-scale took off in the twentieth century, brought about by the development of transport networks. Public transport, such as buses, trains and underground services, together with increased car ownership, meant that people no longer needed to live close to their places of work in the central and inner parts of the city. They could commute over increasing distances.

Development densities in the new suburbs have been so low that they have given rise to what is known as suburban sprawl or 'slurbia'. Its negative effects have recently spawned a new movement in the USA called 'Smart Cities', which aims to curb sprawl and promote a return to compact communities. While the worst excesses of suburbanisation are in North America, towns and cities in other parts of the world show similar trends.

SUBURBANISATION ON THE RAMPAGE

Case study 13

Two American examples

Los Angeles

Los Angeles (LA) is often described as a 'suburban metropolis'. The essential nature of LA's growth is horizontal and sprawling. A population of over 14.5 million lives in an area of 88 000 km^2 — 55 times larger than the area of Greater London. This gives a population density of 165 people per km^2, compared with 4600 people per km^2 in Greater London.

LA is well known for its love affair with the car. This is emphasised by the 1500 km network of roads threading through the sprawling built-up area, the numerous parking lots and ramps, the drive-in establishments of every kind, the wide streets and plentiful shopping centres. LA is also famed for its neatly arranged residential areas, which tend to look the same because of their matching ranch-style houses, complete with outside patios, barbecue grills and swimming pools. The city is a highly polarised society in terms of wealth and poverty, and is still recovering from the race riots of 1992.

Economic success and the influx of people have fuelled the growth of LA (*Case study 10*). The factors encouraging this growth to sprawl and to be of a suburban nature include:
- the coincidence of the city's economic take-off with the arrival of the motor car
- the availability of cheap petrol and land
- high levels of personal affluence
- lax planning controls
- the media's promotion of the suburban dream

However, although many families may have found their dream home, the prevailing suburban lifestyle does have its downsides. These include:
- driving long distances to work or for services and leisure
- congested roads, with slow-moving traffic
- high levels of air pollution caused by road traffic (LA is renowned for its 'killer smogs')

TopFoto

TopFoto

Figure 26
(A) Smog over LA;
(B) Heavy traffic in
Mexico City

Mexico City

With a population estimated at around 20 million, Mexico City contains more people than LA. However, these people are squeezed into an area less than one-thirtieth the size of LA (2500 km²). Since the early 1900s, the population of Mexico City has exploded due to high rates of natural increase and large numbers of rural–urban migrants. All roads seem to lead to Mexico City, thanks to the continuing growth of jobs in the manufacturing and service sectors.

The city, hemmed in by swampy ground on the east and a limited transport system, now sprawls across a former lake bed and into the hills beyond (*Case study 6*). To the north, the current housing boom is expected to add 500 000 residents to just one suburb (Tecamac). There is no end in sight to Mexico City's relentless expansion, which is encouraged by the gradual extension of road and subway systems. The city seems to be continuously on the brink of strangulation and grinding to a halt. Exhaust emissions from the high-density road traffic are trapped by the surrounding mountains. Consequently, air pollution is among the worst in the world, and the health of residents suffers accordingly.

This case study illustrates that to live in the suburbs may be the dream of millions, but it does have a downside. The nature of suburbanisation varies from country to country and, more broadly, from MEDC to NIC and LEDC (Case studies 13–17).

Case study **14** EDGE CITIES

The third wave

Edge cities represent the third wave in the history of suburbanisation. In the first wave, people simply moved their homes out beyond what was traditionally recognised as the city — a high-density urban environment. In the second wave, people tired of having to travel to the CBD for the necessities of life, so the market places were moved to the suburbs. This gave rise to the US shopping malls and the UK edge-of-town shopping centres, particularly during the 1970s and 1980s.

More recently, jobs — our means of creating wealth — have relocated to where we live and shop. This third decentralising wave has led to the rise of edge cities, which are also known by other names, such as suburban business districts, suburban cores, technoburbs and peripheral cities.

The existence of an edge city is usually signposted by a cluster of glistening office blocks, huge shopping complexes and chain hotels. Notable amenities include entertainment, hospitals, schools and sometimes regional airports. The city is usually linked to major roads. There are five rules for a suburban centre to be recognised as an edge city:

(1) It must have more than 464 500 m^2 of office space (about as big as a good-sized downtown district).

(2) It must include over 55 740 m^2 of retail space (the size of a large regional shopping mall).

(3) The population numbers must rise every morning and drop every afternoon (i.e. there are more jobs than homes).

(4) It must be a single-end destination (the centre 'has it all' — entertainment, shopping, recreation etc.)

(5) It must not have been part of a city 30 years ago (i.e. it was largely rural).

Applying these criteria, over 120 locations in the USA are recognised as being edge cities. Of these, 24 are located in the LA metropolitan area, 23 in metropolitan Washington DC and 21 in greater New York City.

Edge cities are particularly attractive as locations for corporate headquarters. With convenient access and pleasant surroundings, edge cities avoid many inner-city problems. However, critics have noted a marked class **social segregation** associated with them, a reduced sense of community and an increase in congestion and crime.

Tysons Corner, just outside Washington DC, is a typical edge city. Located near the junction of main roads, it was not much more than a village a few decades ago. Today, it contains the largest retail area on the east coast south of New York City. This includes six anchor department stores and over 230 stores in all. There are 3400 hotel rooms, more than 100 000 jobs, and over 2 322 500 m^2 of office space. Despite all this, Tysons Corner has no local civic government of its own — its administration is shared by a number of local authorities. Other US examples include Edison (New Jersey), Irvine (California) and Plano (Texas).

There are signs that edge cities might be developing outside the USA. Certainly, developments resembling them are making an appearance in the metropolitan area of Mexico City. In Europe, an Edge Cities Network was initiated by the London Borough of Croydon in 1996. Its partner members now include Loures (on the edge of Lisbon), Getafe (Madrid), Nacka (Stockholm), Espoo (Helsinki), Ballerup (Copenhagen), Fingal (Dublin) and North Down (Belfast).

Edge cities are not considered to be part of what is often called the post-suburban era. They are seen as an integral part of evolving suburbanisation — the product of its third and most recent wave. Edge cities might also be seen as instruments of suburban intensification.

Question

Draw an annotated diagram to illustrate the three waves of sub-urbanisation.

Guidance

You could show the situation as a simple diagrammatic section (as in Figure 27) or in plan view, perhaps using a scheme of concentric zones (as in Figures 33–36, pages 58–59).

All three suburban waves have led to three other processes — **filtering**, **gentrification** and **exclusion**. Each of these processes has made its mark on the social map of the city.

Case study **15** FILTERING AND GENTRIFICATION

Sorting people

Filtering is the process by which housing passes down from higher- to lower-income occupants (Figure 27, p. 47). It is a component of suburbanisation. The addition of each new generation of suburb is generally spearheaded by more wealthy people. As they move out to their smart new homes, they leave behind houses that become occupied by families that are not so affluent. When these families move, they, in turn, vacate properties for even less well-off people. In this way, the chance to 'upgrade' in terms of housing gradually 'filters' down in a wave-like manner from rich to poor. At the end of the filtering chain are properties abandoned by even the poorest people. Abandonment means the creation of brownfield sites ripe for redevelopment.

Some of the momentum for filtering is provided by buildings becoming obsolete over time. Physical deterioration inevitably lowers property values and the general perception of neighbourhoods. The result is the creation of housing opportunities for poorer people.

Clearly, the process of filtering encourages the segregation of people on the basis of wealth and socio-economic class. This process is also active in LEDC and NIC cities.

Gentrification is filtering in reverse. It denotes a significant neighbourhood change in many cities, involving the improvement and upgrading of older properties in inner-city areas, the movement out of low-income tenants and the in-flow of high-income owner-occupiers. The revival of Washington DC's Georgetown district during the 1960s is a well-known, early example of gentrification, so too is the upgrading of Islington in inner London. In both cases, the process involved young couples or single people moving into the area, who were usually well-educated and professional by employment. However, also present in these gentrified areas were more affluent middle-class families, previously living in the outer city. This implies that significant numbers of people were choosing to live in the inner city. They were reacting against first- and second-wave suburbia.

The main attractions of inner-city gentrification are:
- easy access to the city centre and its amenities
- short journeys to work and savings on commuting costs
- properties with large rooms and grants to help with improvement
- distinctive houses as opposed to ubiquitous suburban estates

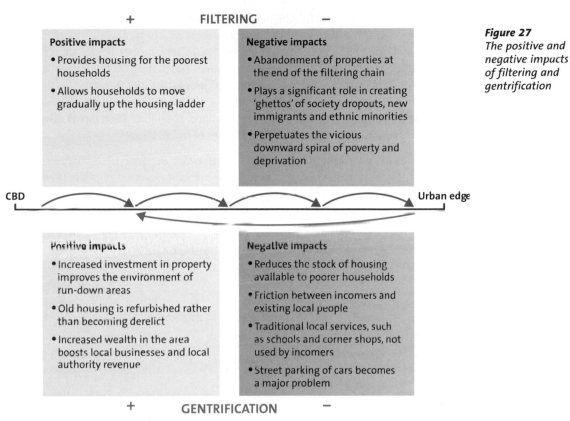

Figure 27
The positive and negative impacts of filtering and gentrification

Figure 27 shows the positive and negative impacts of gentrification.

Falling somewhere between these filtering and gentrification processes is studentification. This is the term given to the social and environmental changes caused by large numbers of students living in particular areas of a town or city. It involves converting single-family residences into multi-occupant houses. As the student population becomes concentrated, there follows an impact on local retailing in the form of take-away food outlets and cheap off-licences.

Studentification is the outcome of a massive expansion of higher education opportunities in the UK and other MEDCs. There is nothing new about the process — cities such as Oxford, and Heidelberg in Germany, have long been renowned for their student populations. What seems to be new is the degree of spatial concentration in particular parts of the built-up area. Manchester today is considered to be the 'student capital' of Europe with concentrations in districts such as Fallowfield.

Studentification has a positive outcome, in that it helps to regenerate poorer areas of cities. It could almost be seen as a particular form of gentrification. The housing stock is improved and, as a result of spending their grants and loans, students help to boost the local urban economy. They also contribute to the creative aspect of the city, for example, with music and the visual arts. In cities suffering deindustrialisation, such as Leeds and Newcastle, students who stay on after graduation help to stem out-migration flows, stabilising the adult urban population. Cities such as Cambridge and Bristol in the UK, and Boston and Berkeley in the USA, have benefited from their long-term association with higher education. There have been spin-offs in the form of science parks, business units, research and development institutions and incubator units.

Although there are these positive aspects of studentification, there are also negative effects. The arrival of increasing numbers of students pushes up property values as landlords cash in or parents acquire buy-to-rent properties for their children. Environmentally, the transient nature of the student population often leads to less pride being taken in maintaining the appearances of properties and gardens. Furthermore, students are concentrated in a relatively narrow and young age band, so their lifestyle often clashes with that of the older populations found in such areas. Dwindling numbers of families can also lead to the closure of local schools.

The processes examined in this case study lead to either the general decline or improvement of city areas. Each process has both positive and negative impacts. Remember, improvement in one area might well prompt decline in another, and vice versa.

13 Question

Using Figure 27 as a template, draw a diagram to summarise the positive and negative aspects of studentification.

Guidance

Most aspects are mentioned in the text but you might do some role playing — as a local resident and as a student — to reveal some more pluses and minuses.

Website: **www.geographyinthenews.rgs.org/news/article/**

Another process that helps to mould the social map of the city is **exclusion**. This involves barring particular groups of people from living in certain parts of the city. As a consequence, those excluded groups become concentrated in another part of the city. Exclusion is not a new process. The **ghetto** is a long-standing outcome of this process, from the Jewish ghettos of European cities in the Middle Ages to the black ghettos of US cities in the twentieth century, such as Harlem in New York. The ethnic segregation of the ghetto is the outcome of internal and external factors (Figure 45, p. 72). Exclusion, like filtering and gentrification, plays its part in the spatial sorting of different groups of people. Gated communities may be the most recent and blatant examples of exclusion.

Case study 16   EXCLUSION

The world of gated communities

A **gated community** is an area of wealthy, private housing, with a secure perimeter wall or fence and controlled entrances for residents, visitors and their cars. Round-the-clock security patrols add to the sense of protection. Many gated communities contain a range of services and amenities, making them more than just dormitory enclaves. They are often homogenous communities in terms of social class, race and culture. Gated communities are, therefore, an extreme form of segregation and exclusion. They are the product of income inequalities and a growing concern about personal security. Clearly, they are increasingly important influences on the social and cultural maps of cities.

There is a long history of gated communities — from royal palaces, monasteries and guilds in medieval cities in Europe, to oil-worker compounds of Middle Eastern cities today. The 2000 census revealed that 7 million US households (6% of the total population) lived

Using case studies

in developments defined by walls and fences, and controlled by gates requiring entry codes, key cards or scrutiny by security guards. Gated communities are more commonly found in 'sunbelt metro' areas such as Los Angeles, Dallas and Houston, but their popularity is spreading to places as far apart as Long Island (New York), Chicago, Atlanta and Washington DC.

There are now more than 1000 gated communities in England alone. These range in size from a small cluster of new houses in a village or town, to blocks of flats, such as in Bow Quarter, east London. Opinion is divided about such communities. The police view them as increasing tension, and describe difficulties in controlling surrounding areas. On the other hand, the government supports the idea and says they have an important role to play in urban renewal programmes and in creating crime-free areas. They may also help meet the special needs of elderly people, who seek a secure environment and sheltered accommodation (Case study 25).

Gated communities are also springing up in newly industrialising countries (NICs) and rapidly industrialising countries (RICs). Examples range from the Oriental Grand Garden in the Pudong area of Shanghai (Case study 18) to the six compounds in the Tamborei suburb of São Paulo. In Johannesburg, South Africa, which has one of the highest crime rates in the world (16 times higher than London), the number of gated communities is increasing (Case study 20). These communities are mainly for the benefit of the wealthy white population. To some, they appear to be a return to the old days of apartheid.

In India, software export and service outsourcing are creating a rising middle class, leading to a boom in the demand for gated housing. A large Indian company has just embarked on a programme to provide 800 000 gated homes across the country. In addition to boundary walls, armed guards and 24-hour security, these communities have their own private fire service, back-up electricity generators, water filtration plants, schools, cinemas and green spaces. There are no cows, beggars or litter. Residence in such a gated community is an outward sign of personal success, 'Bollywood' style. Low-interest loans have made possible something financially and culturally unthinkable a few years ago — borrowing money to buy homes, cars, luxury goods and foreign holidays. Some architects and planners fear the rising stock of gated communities is increasing polarisation in Indian society. Although their design is Western in appearance, they are effectively reinforcing the Hindu caste system, with its in-built social separation.

Today's gated communities are stark expressions of exclusion. Many have been built for reasons of personal security and safety. Within the highly polarised populations of modern cities, is it the rich who fear the poor?

14 Question

Carry out some research into the ways in which the black ghettos of US cities differ from gated communities.

Guidance

Websites: www.boston.com/news/globe/ideas/articles/2004/01/25/inside_the_gates
http://en.wikipedia.org/wiki/Ghetto

Using case studies

The emphasis in this section has been on suburbanisation and its impact on MEDC cities. The following case study summarises the situation in the world of LEDCs.

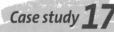
Six city thumbnails

Mexico City (Mexico)

See *Case study 13* for the suburban nightmare that exists here.

CHALLENGES

- More control and planning
- Improvement of housing, including proper servicing
- Better transport
- Reduced pollution
- Job generation

Shanghai (China)

This boom city seems to be combining the characteristics of a sprawling Los Angeles and a densely packed Manhattan. There are cores of high-density development dotted like 'islands' in a suburban 'sea', forming the Shanghai–Nanjing–Hangzhou corridor. This huge urban creation is held together by some 3000 km of elevated subway. Suburban growth is managed by a highly centralised planning system, unlike in many other parts of the world, where the individual cities control their own destinies. Increased well-being and personal mobility are leading to more affluent households living in more peripheral locations.

Baghdad (Iraq)

The Iraq capital is, by Arab standards, a spread-out, low-rise city. This is encouraged by the fact that the Mesopotamian Plain presents few physical barriers, except rivers. Thus, the suburbs sprawl along both banks of the River Tigris and westwards towards the River Euphrates. In the 1960s, most of the city's many shanty towns were pulled down and a new city for the squatters was built on the east bank of the Tigris. Much of Baghdad is a Shi'ah stronghold. Suburban life is being seriously disrupted by the bombings that are part of the present civil unrest. Many people live in fear of their lives.

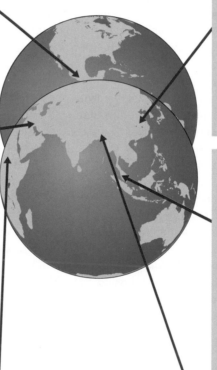

Kuala Lumpur (Malaysia)

Unlike most Asian cities, the central area is surrounded by a belt of relatively low-density suburbia. Suburban sprawl has been checked by the creation of a poly-nucleated pattern of new towns and smaller suburban centres, located along the major arterial routes of the city. Terraced housing developments in established middle-income commuter suburbs have been booming, especially since the building of the Putra Light Railway Transit system that links the eastern and western suburbs of the city.

Addis Ababa (Ethiopia)

The distribution of wealth in this city follows more closely the North American/European model. Poverty and slum housing are most intense in the neighbourhoods surrounding the CBD. Government officials, wealthy businessmen and elite foreigners live in spacious compounds (LEDC equivalents of gated communities) far away from the noise and congestion. Such compounds are found in two concentrations, each close to one of the capital's airports. High rates of natural increase and in-migration are causing an unchecked mushrooming of the suburbs along the main roads leading into the city.

In contrast to MEDC cities, the suburbs of LEDC cities are mostly poor areas. There is little or no regulation, so areas of slum housing (shanty towns) can spring up almost anywhere, particularly on land avoided by more wealthy people. The two extremes of housing are often found close together.

Dhaka (Bangladesh)

Huge shanty towns exist in suburban locations around the city centre. However, a growing middle class is showing a preference for suburban location. This is inflating property prices and generating boom-type expansions on both the northern and southern fringes of the city. The lack of effective planning means that many of these suburban expansions are being built in areas where there is a serious flood risk.

Figure 28 Recent events in LEDC and NIC suburbs

Question

Figure 28 lists five challenges for LEDC and NIC suburbs. Read the six case studies summarised in the diagram again. Which do you think is the greatest challenge? Give your reasons.

Case study 15 looked at the process of gentrification. This is part of a broader process of change, commonly referred to as **regeneration**. Gentrification is about housing and social class. By comparison, regeneration relates to all aspects of the urban fabric and economy. Regeneration often goes hand-in-glove with **re-imaging** in attempts to turn around the fortunes of urban areas, and sometimes even whole towns and cities. **Reurbanisation** is another term used to identify these processes aimed at putting new life back into abandoned or under-used parts of the city. Reurbanisation is best illustrated after we have some understanding of another, but external process, known as **economic globalisation**.

The nature and impact of economic globalisation

Economic globalisation, post-industrialism, demographic change and government intervention are four external processes of considerable consequence, in terms of their impact on the growth and structure of cities. They are all aspects of economic development. Together, these factors provide the basis for urban growth. They are currently responsible for much of the dynamism in the global urban system. We shall take a closer look at two of these processes: economic globalisation and post-industrialism.

Globalisation is any process of change operating at a world scale and having worldwide effects. It can be physical (e.g. sea-level change), human (e.g. economic development) or both (e.g. global warming). Economic globalisation is related to the growth of the global economy, and the fact that the production of goods and services is becoming increasingly footloose. Firms today can locate almost anywhere in the world. This has been made possible by advances in transport and communication. Organisations are able to take full advantage of least-cost locations. They are free to exploit pools of cheap labour and other comparative advantages. Not only is economic production spreading across the globe, but so too is consumption. Brand-name goods are now both available and in demand throughout the world. Globalisation is also apparent in urban planning and management — for example, in international exchanges of strategies and best practices.

The big players in economic globalisation are TNCs. They are the main decision-makers in changing the location of production. Through their advertising and promotional campaigns, they are able to manipulate both demand and consumption.

The effects of economic globalisation are being felt most in LEDC cities. Firms are being drawn to them by lower labour costs and their pools of educated and skilled workers. The rapid urbanisation being experienced in so many LEDCs is creating expanding markets for goods and services. Many LEDCs are prepared to reduce restrictions, such as health and safety laws or planning regulations, in order to attract **inward investment**. Tax incentives and subsidies are also being offered.

Case study 18 SHANGHAI

A beneficiary of economic globalisation

It may be surprising to find that one of the cities benefiting most from economic globalisation is located in one of the few remaining communist countries in the world (*Case study 5*). Why is China opening its doors to market forces and the global economy? Why single out Shanghai?

The answer to the first question is that China is keen to become one of the world's most powerful economies. On present trends, it is possible that the country will rank second to the USA in about 10 years' time. The wish for fast economic development is related to China's huge population. On one hand, there is a need to generate as much employment as possible. At the same time, thanks to its present cheapness, the huge workforce is perhaps China's greatest resource and comparative advantage. If China is to achieve fast economic growth, it needs to attract inward investment and expertise, but these could destabilise China's communist regime. Therefore, China took the decision in 1990 to let in foreign companies, but only to a limited number of clearly contained locations. The most important of these locations is Shanghai. The thinking behind the decision is that a 'globalised' Shanghai will act as a catalyst of economic development elsewhere in China.

Figure 29
Key features of the Shanghai municipality

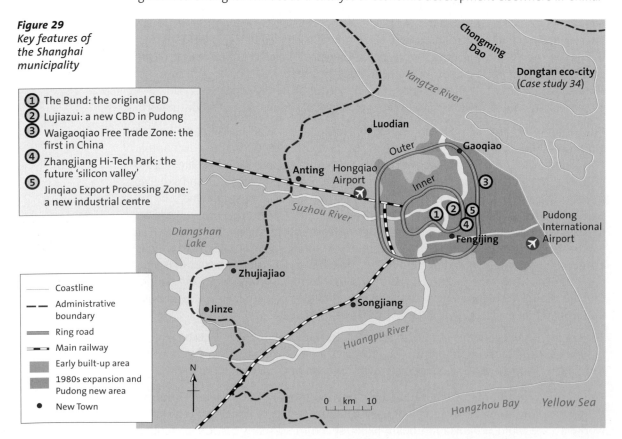

① The Bund: the original CBD
② Lujiazui: a new CBD in Pudong
③ Waigaoqiao Free Trade Zone: the first in China
④ Zhangjiang Hi-Tech Park: the future 'silicon valley'
⑤ Jinqiao Export Processing Zone: a new industrial centre

— Coastline
– – Administrative boundary
═══ Ring road
━•━ Main railway
▨ Early built-up area
▨ 1980s expansion and Pudong new area
● New Town

0   km   10

Contemporary Case Studies

The question is: why Shanghai rather than any other leading Chinese city?

- Shanghai is the largest city in China.
- It has a long history as a gateway to foreign trade.
- It accounts for 25% of the country's export earnings.
- During the communist era, it has become the country's most important centre for heavy industry.
- It is within a 2-hour flight of Japan and key Asian Tiger cities such as Seoul, Taipei, Hong Kong and Bangkok.

In order to aid Shanghai's participation in economic globalisation, a major economic growth zone was created at Pudong, a greenfield site on the east of Shanghai (see Figure 29 and Case study 29). From being a remote and unhealthy backwater of the Yangtze estuary, Pudong has become the smartest and most desirable part of Shanghai in which to live and work. By 2002, it was already accommodating a population of almost 2 million, nearly 200 high-rise buildings had been constructed for office use and overseas investors from over 80 countries had interests there. Foreign investors in Shanghai enjoy preferential tax concessions.

Much of the current foreign investment is focused on the high-tech park containing 'incubator' enterprises, research and development institutions, and ICT companies. The aim is for Shanghai to become another 'silicon valley'. Other significant components of Pudong include export processing, free-trade and financial zones. The financial zone has already become Shanghai's main CBD. Nearly 150 local and foreign financial institutions are now located there, along with the Shanghai Stock Exchange.

In a short period of time, and with the support of the government, Shanghai has become China's flagship city and an important node in the emerging global economy. The city has been a major beneficiary of economic globalisation.

Much of the global production of goods and services increasingly found in LEDC cities has come from MEDC cities — once the bastions of the industrial world. This global shift has created an urban crisis in MEDCs. Many cities are the victims of economic globalisation. In order to survive, they have had to find new economic roles to compensate for their loss of manufacturing and industries (i.e. deindustrialisation). They have had to face a new era and a new set of circumstances, collectively described as post-industrial. Possible options to be considered include:

- attracting new types of manufacturing, particularly those at the cutting edge of new technology — for example, in the fields of biotechnology and telecommunications
- developing a reputation in the tertiary sector — for example, for sports and leisure facilities or for culture
- providing specialist services in the quaternary sector — for example, computer programming, software design, advertising, and research and development

Cities suffering from deindustrialisation require 'repackaging'. They need to cultivate a new image that will attract investment, business and people. The areas scarred by industry and related activities have to be transformed and given new uses. The key requirements are regeneration and re-imaging. A beacon venture in the UK was the London Docklands project. It showed what was possible by way of reurbanisation, albeit in just one, relatively small, part of a **world city**.

The cases of Bradford and Southampton

Bradford

With a population of over 450 000, Bradford in Yorkshire is one of the ten largest cities in Britain. Buildings like the City Hall, the Wool Exchange and the Manningham Mills chimney are reminders of the city's early growth, which was fuelled by the wool and textile trade. The city is famous for its links with the Brontë sisters and the artist David Hockney. It is also renowned for the race riots of 2001. The city has been keen to shake off this last image by embarking on a multi-million pound regeneration scheme.

The city's economic base, culture and society have completely changed following the decline of textile manufacturing. There has been a serious decay in its physical fabric, accompanied by social exclusion. The symptoms of social deprivation are plain to see. Parts of Bradford are among the poorest areas in England, being sixth worst for unemployment (4.6% compared with the national average of 2.8%) and fifth worst for low incomes (44% of children live in low-income households compared with the national average of 27%). In 2000, only 34% of pupils achieved five or more GCSEs at A* to C grades compared with the national average of 50%. Recruiting for skilled technical, professional and managerial posts is difficult.

A large Muslim population plays a significant part in the city's business and cultural life. Immigrants from south Asia were attracted here in the third quarter of the twentieth century by jobs in the textile industry. Pakistani and Bangladeshi communities make up a fifth of the city's population.

Housing in one of Britain's most multicultural cities is becoming increasingly segregated. The more affluent minority families are too scared to move on to the better, mostly white, estates in Bradford. Some estate agents have tried to encourage the movement of white families into Asian areas of the city. This move has been blamed for the serious outburst of rioting in the city's Lidget Green area in 2001, when Asian youths reacted to rumours that white outsiders were 'invading their territory'. A large proportion of the old housing stock is substandard.

The challenge for Bradford is to erase its negative image and to exploit a number of its attributes positively: multiculturalism, tourism, modern industries and services. To do this requires investment and action in three key aspects of the city: the environment, the economy, and society.

Regeneration is essential if Bradford is to reduce its environmental and social problems. This will require hard measures

www.flickr.com/photos/johnleach

Photo courtesy of Bradford Council©

Figure 30 *Bradford: old and new*

(demolishing or repairing rundown property) and soft measures (confidence-building, education and providing better opportunities for employment). Heritage tourism is being encouraged as a catalyst for change. Many factories of the Industrial Revolution have become museums, craft centres and galleries, or divided to provide small business units. Warehouse conversions are increasingly popular. Saltaire, famous as a Victorian model village created by Titus Salt for his factory workers, has been designated a UNESCO World Heritage Site, giving it the same status as the Taj Mahal and the Pyramids of Egypt.

A scheme called 'Vision of Bradford in 2020' has recently been implemented. It is based on a 'park in the city' concept that offers open green and leisure spaces, re-introduces water into the city centre and exposes many of the city's heritage of listed buildings. The scheme involves four regenerated quarters within the heart of the city. Each will create new spaces for commerce, education and leisure, as well as incorporating the natural feature of the River Aire, long buried under the city.

Economic diversification requires that the occupational structure shifts more towards modern lines of manufacturing (engineering and chemicals), service industries, telecommunications, information technology, leisure and tourism. The Integrated Development Plan helps to generate investment in the city and links this with job creation and training for local people. With 8 million visitors a year, Bradford is becoming an important tourist destination. Among its attractions are the National Museum of Photography, Film and Television and a range of annual festivals.

Perhaps the greatest challenge for Bradford is to improve the quality of life of many of its residents. The provision of better services (shops, schools, medical centres and leisure facilities) and more employment opportunities are two vital steps in this direction. Better housing is the third priority. The inner city and some of the worst social housing estates are scheduled for redevelopment or improvement. There will be a big push to provide more social housing.

The issue is whether all the regeneration and re-imaging funding can really tackle the fundamental problems of Bradford's historical legacy. Regeneration is about more than just bricks and mortar. The goals for any city regeneration project are to breathe new life into the area and to create an environment that supports the people who live and work there.

Southampton

Southampton has a worldwide reputation as a port city. This fame is founded on the great ocean liners that carried huge numbers of passengers across the Atlantic, and to and from British colonies in Africa, Asia and Australasia, during the nineteenth and twentieth centuries. In order to remain in business, the city has had to make regular changes to its port traffic. When air transport reduced ocean-going passenger traffic, the port turned first to the cross-Channel ferry business and then to containerised cargo. Competition from nearby Portsmouth meant there was little success with the passenger ferry business, but the container side has boomed. However, this is now under threat because the port has run out of waterfront space (*Case study 12*). The cruise business is another flourishing aspect of the port business.

Space has been a challenge for other aspects of the city's livelihood. For decades, the spread of the built-up area has been limited by a green belt. Suburbanisation has now led to the spawning of detached suburbs beyond it (Figure 24). With no vacant sites inside the green belt, there has been a fear that the city might lose its status as a major retail centre. Southampton has been left with no option other than to revamp its central area and maintain a strong centre, which has been undertaken successfully. This has involved three flagship developments (Figure 31):

- Ocean Village — a large marina development with housing and leisure facilities
- Southampton Oceanography Centre — a world-leading ocean research institution
- West Quay Retail and Leisure Park — one of the largest projects of its kind in Europe

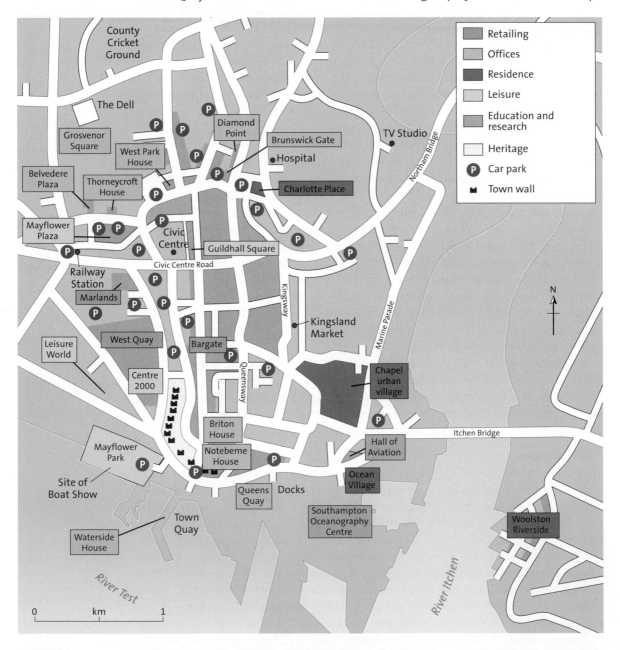

Figure 31
Regeneration in central Southampton

On the other side of the River Itchen, another flagship scheme, called Woolston Riverside (Figure 32), has just been set up. This development will be built on a 12-hectare brownfield site. It will transform a redundant shipbuilding facility into a vibrant, thriving and sustainable, mixed-use waterfront area, with a marine business park, high-quality housing, and retail, leisure and community facilities. It will also contribute positively to the physical, economic and social regeneration of the surrounding area, a rather rundown early suburb of the city.

Regeneration has been a vital process in both cities. In the case of Southampton, it has been a matter of image polishing, rather than re-imaging. In this respect, the modern port contrasts with Bradford, which has had to embark on a much more fundamental 'facelift'.

Figure 32 *Computer-generated image of the proposed development of Woolston Riverside, Southampton*

16 · Using case studies

Question

Write a short essay about the different circumstances that cause cities to embark on re-imaging projects.

Guidance

Read *Case studies 18* and *19* and take a look at *Case study 33*.

Models

We now return to the subject of models. Virtually all cities show a degree of concentric zoning because cities inevitably grow outwards from a historical nucleus. Hence, it is possible to recognise in all cities, no matter where they are located in the world, four recurring components:
- a core
- an inner city belt
- a suburban ring
- a rural fringe

We may add two more 'global' generalisations about the city pattern, moving outwards from the core, namely that:
- the general age of the built-up area decreases
- the overall density of development tends to decrease, due to the expanding perimeter of the urban fringe and the impact of transport improvements

Clearly, the relative extent of the first three components of the urban pattern will vary greatly from city to city. Key influences include the age of the city, the speed and dynamism of city growth at any particular time, the availability of land and the efficiency of transport. Equally, these concentric zones are readily distorted and even truncated by the physical geography of the individual city. The distorting factors include:
- **Waterfronts** — coastal, estuarine and riverine — because they attract particular types of land use and sometimes housing; the zones end at the waterfront.

- **Transport routes** — valleys tend to be exploited by transport routes linking to other cities. Because of their enhanced accessibility, they become the lines of least resistance to urban growth. The radial routes leading from a city centre tend to create sectors and distort the concentric zones into a cobweb pattern.

Why is it not possible to produce one model to fit all cities? The reason is that the processes, both within and beyond a city, vary significantly, depending on the general level of development of the country in which the city is located (Figure 15, p. 24). Different processes produce different outcomes. There are other distorting factors (of a human kind), such as the degree of government intervention through planning, cultural traditions and the course of history.

Figures 33–36 show models summarising the essential character of a typical city at four different points along the development pathway. Each figure shows the same quadrant of four zones, but the precise character of these zones varies because of differences in processes and the availability of resources. At best, these four models provide summary snapshots of today's dynamic cities.

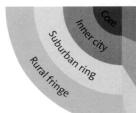

FEATURES AND PROCESSES
CBD historical nucleus and therefore some heritage in the form of colonial buildings; wealthy residents in gated communities of high-rise apartments; exclusive shops and services
Original shanty towns at inner margins, now subject to consolidation and improvement, including provision of basic services; workshop-type industry and services; some middle-class homes in older properties
Segmental mix of housing types, including owner-occupied apartments and single-storey dwellings; government low-cost housing schemes of high-density housing and basic amenities; most recent shanty towns towards the outer margins on low-cost land — high density with low service provision; manufacturing and light service industry on purpose-built estates along main roads and near airports; some signs of TNCs
A few new satellite towns built to relieve pressure on the city and attract foreign investment; some commuting on a seasonal or part-time basis to supplement farm income; local agriculture geared to the nearby urban market; specific locations of high scenic or amenity value becoming exclusive gated enclaves for the most affluent; early signs of decentralisation, but otherwise still overwhelmingly rural

Figure 33 Model of an LEDC city

FEATURES AND PROCESSES
CBD undergoing expansion and redevelopment; strong presence of TNC offices; cloned skyscraper cityscape; some of the historical nucleus remains; cramped artisan housing being refurbished as part of heritage promotion; ex-colonial quarter of grand residences, parks and wide roads; home to majority of remaining ex-patriots
High-density mixture of poor housing and industry; inner margins being redeveloped to provide space for CBD expansion; tracts of government-financed housing; infilling by early shanty towns on land avoided by formal development
Lower-density development but apartment blocks much in evidence; segregation of housing and industrial estates; increasing provision of low-order services; planned, often gated, communities for rising middle class, often adjacent to high-tech parks with strong TNC presence; new shanty towns along the outer margins
Rural dilution underway; rising numbers of commuters from outlying settlements; rural population making increasing use of city services; city population making greater use of rural leisure amenities

Figure 34 Model of an NIC city

FEATURES AND PROCESSES	

Re-imaged core; CBD undergoing revival in emerging market economy; appearance of Western chain stores and fast-food outlets; historical nucleus — churches, palaces and other heritage buildings refurbished, mainly to attract foreign tourists and hard currency; relics of the socialist era include large open spaces for demonstrations and rallies, massive sculptures symbolising the struggles of the proletariat, 'wedding cake' city hall

Former middle- and upper-class housing of the pre-socialist era close to core now being refurbished and occupied by those who have 'made it' during the transition to a market economy; mafia money behind much of this improvement; increasing evidence of wealthy class; areas of former working-class housing towards the outer margins also showing signs of improvement

Residential districts built during the socialist period within a sort of greenbelt framework; poorly built, badly maintained high-rise apartment blocks — many seriously dilapidated; each district has its own welfare and basic consumer services; industrial areas located between the residential districts in order to minimise commuting; a belt of post-socialist development emerging in the form of low-rise apartment blocks at the outer margins — new functions beginning to emerge — hotels, restaurants and supermarkets

Much evidence of neglect in the agricultural landscape — huge farm buildings lie abandoned since the collapse of the collectives; money made in the city is being invested in the improvement of the stock of dachas (country second homes) — part of a move towards more rural leisure

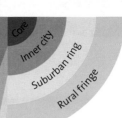

Figure 35 Model of a post-socialist city

FEATURES AND PROCESSES	

CBD fighting competition from edge-city developments, re-imaging and redevelopment; increasing role of leisure in the promotion of the '24-hour' city; segregation of central businesses into specialised areas; conservation of heritage buildings; zones of assimilation and discard are evident towards the outer margins

Originally much of the belt occupied by a mix of industry and working-class housing; substandard terraced housing replaced by 1960s high-rise apartment blocks; filtering has led to increasing concentrations of poorer households, ethnic minorities and students; deindustrialisation has left a heritage of brownfield sites — some being used for the construction of expensive housing for the young upwardly mobile; towards the outer margins, lower-density housing created during the boom in public transport at the turn of the nineteenth century now increasingly middle class; gentrification much in evidence

Mainly low-density residential development facilitated by public transport and increased car ownership; large areas of owner-occupied housing; estates of social housing serving the needs of poorer households; neighbourhood service centres meet needs of local residents; recent signs of intensification and general raising of development densities; purpose-built estates of light and service industry at the margins of the built-up area; large leisure and retail parks, together with hospitals, schools, sports stadia and recycling centres

Outward spread of urban fringe contained by greenbelt and strategic gap policies; provision for recreation within the greenbelt; protection of agriculture and conservation of villages, but pressure to use space for infrastructure services (water, sewage, power generation); beyond the greenbelt, villages increasingly urbanised as they become dormitory or commuter settlements; new and expanded towns are evidence of planned decentralisation

Figure 36 Model of an MEDC city

17 **Using case studies** **Question**

Check the patterns and processes of Addis Ababa (*Case studies 4* and *26*), St Petersburg (*Case study 9*), Shanghai (*Case study 18*) and London (*Case study 11*) against the appropriate model in Figures 33–36. How closely do they match?

Part 5

Urban issues

Despite the advantages they offer, towns and cities are not perfect places in which to live and work. There is a downside, made up of a range of problems or issues. Housing, congestion and pollution are three examples. Many of the issues arise because the growth of cities and the spread of urbanisation are spontaneous and random. It is often not until an area has become built up that people become aware of its imperfections and problems. Once this happens, people look to the decision makers, particularly government ministers and planners, to resolve these issues. Intervention is largely reactive and remedial, rather than proactive.

City issues are often tackled by public–private partnerships. However, you will see in part 6 that there have been instances when individuals or groups of people have tried to put into practice their own visions of an ideal urban future. They have tried to manage things in a more positive way, rather than clearing up the unwanted outcomes of urbanisation.

In this part of the book, the main topics are:
- identifying the range of major city issues that exist today
- finding out whether the nature and mix of issues change as cities and countries move along the urbanisation pathway
- looking more closely at recent issues that are less well covered in A-level textbooks

Changing city issues

The issues facing cities around the world today are diverse in character and scale. They range from the universal challenge of providing decent housing for the poor, to much more localised issues, such as what to do with a brownfield site or how a city can change its image.

These issues can be classified in various ways. Perhaps the simplest is to place them under one of the five broad headings shown in Figure 37. However, this is not entirely satisfactory, as many issues are multi-dimensional. For example, traffic congestion is more than just an environmental issue. In addition to being a prime cause of environmental pollution, it creates economic costs and has socio-cultural impacts through its effects on the quality of life, particularly health. It is also an issue that often attracts government intervention.

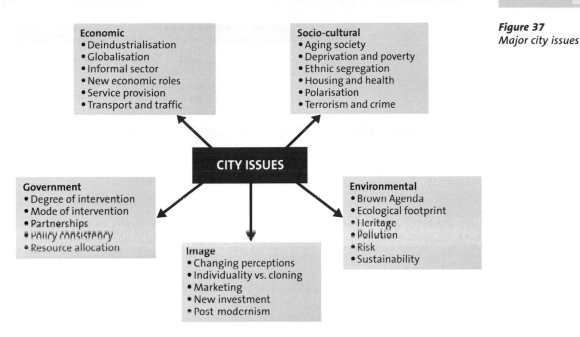

Figure 37
Major city issues

Figure 38 illustrates how city issues can be grouped into another set of broad categories — in this case, what happens along the urbanisation and development pathways:

- issues that decline in relative significance (e.g. poverty and deprivation)
- issues that become more significant (e.g. the **ecological footprint**)
- issues that remain fairly constant (e.g. traffic congestion)

The significance and scale of issues, as well as the actual mix, change along the pathway. This is best illustrated if we compare the issues confronting post-industrial cities, such as London, with those facing industrialising cities, such as Mexico City or Shanghai. Some issues are common to both scenarios, while others are not:

- **Post-industrial cities:**
  - promoting the new economic role
  - reusing brownfield sites
  - enlarging the housing stock to meet social and cultural shifts (e.g. more single-person households)
  - meeting rising demands for leisure and retailing
  - dealing with the legacy of an old and inefficient infrastructure (e.g. water supply, sewage disposal and road systems)
  - fighting crime and terrorism
  - providing better social services
- **Industrialising cities:**
  - promoting the new economic role
  - attracting inward investment in a competitive, globalised world
  - reducing environmental impacts (Brown Agenda)
  - enlarging the housing stock
  - combating poverty
  - providing the necessary physical infrastructure (transport)
  - meeting the needs of a growing middle class; tightening up on regulation

SELECTED CITY ISSUES		URBAN PATHWAY AND DEVELOPMENT GROUPINGS				
		Low income	Lower-middle income	Upper-middle income	High income	
		Ethiopia (Addis Ababa)	China (Shanghai)	Mexico (Mexico City)	USA (Los Angeles, Chattanooga)	UK (London, Southampton)
		Bangladesh (Dhaka)	Iraq (Baghdad)	Russia (St Petersburg)	Singapore	
Economic	**Nature of issue**	**Scale of problem**				
Globalisation	City growth today, particularly in LEDCs, hinges on becoming involved in the global economy					
Retailing	Attracting increasingly affluent customers and improving accessibility; making shopping more of a leisure activity; cloning					
Re-imaging	Need to promote a modern, post-industrial image; making the most of the best aspects of heritage					
Social						
Poverty and deprivation	Breaking the cycle of poverty; poverty is never eliminated — its incidence is reduced					
Housing	Improving basic living conditions and access to decent housing					
Terrorism	Safeguarding security and rooting out those who threaten both lives and quality of life					
Environmental						
Risk	LEDC and MEDC cities alike are at risk from natural hazards					
Traffic congestion	All cities suffer congestion and the pollution costs of transport					
Ecological footprint	As cities and the complexity of their needs grow, so does their impact on resources and the environment					

Figure 38 *City issues and the urbanisation pathway*

The rest of part 5 takes a closer look at a number of issues currently confronting cities in different parts of the world. These issues fall under the five headings shown in Figure 37, some of which have already been explored in part 4. We will now focus on a selection of issues that are recent, topical and less well-covered in other textbooks.

Economic issues

Case studies 20 and *21* deal with economic issues that are most keenly felt in LEDC cities. They involve different spatial scales: local, national and regional. *Case study 22* is linked to the affluence and high levels of consumer spending that characterise most MEDCs.

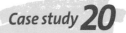

| MISSING OUT ON GLOBALISATION | *Case study* **20** |

South Africa and Johannesburg

Globalisation has largely passed Africa by, mainly because its countries lack the infrastructure, human resource base, formal economy and strong, stable government policies needed to attract inward investment. The continent does not contain a single NIC. However, South Africa is rapidly moving along the development pathway, and is beginning to emerge as a contender for NIC status. Its main city, Johannesburg, is actively trying to achieve global city status.

Globalisation has led to cities everywhere reassessing their ability to promote trade and attract new investment and technology. In South Africa, the former apartheid policy isolated the country from much of the world. Since 1994, however, the government has actively encouraged economic globalisation. South Africa now has a relatively balanced hierarchy of cities and towns. It possesses a higher level of resources than other African countries and is better placed to cope with a range of urban challenges. It has one mega-city in the Gauteng urban region, which is a combination of Johannesburg, Pretoria and the East and West Rand.

Walter Knirr© City of Johannesburg

TopFoto

Figure 39 *Contrasts in Johannesburg*

Johannesburg boasts many of the trappings of globalisation: a CBD with high-rise buildings, 5-star hotels, clean-air industrial estates powered by electricity, and a range of comfortable, luxurious housing. There are branded foods, clothes, sports goods and CDs in the shops. To the southwest of the CBD, the sprawling residential townships are effectively suburban ghettos. They include the infamous Soweto, with a population of nearly 1 million black people. This projects a rather different image of the city. How easily does the scale of urban poverty and exclusion equate with the notion of a global city? In 1886, the site of Johannesburg was farm land. By 1936, after the discovery of gold, the city's population had risen to 475 000. The city has since grown as a service centre for the Vaal Triangle industrial area and its CBD generates 12% of South Africa's GNP. It looks likely that Johannesburg may soon become Africa's first global city.

South Africa is located at the 'remote' end of the world's least developed and least urbanised continent — a serious drawback. This case study shows that such obstacles can be overcome, provided there are resources, stable government, enterprise and a will to succeed. In order to get on the economic globalisation bandwagon, it is vital that a city projects the right image to attract inward investment.

18 Question

Suggest reasons why Shanghai (*Case study 18*) appears to be more successful than Johannesburg (*Case study 20*) in its bid to become a global city.

Guidance

Start by thinking in terms of location, image and government direction.

Case study 21 — THE INFORMAL ECONOMY

Keeping the city going

One in five people in the world struggles to live on less than US$2 per day. In countries such as Bangladesh and Ethiopia, underemployment is a consequence of a surplus of labour. People work less than full-time, even though they are willing and able to work more hours. Most are forced to find other ways of making a living outside the formal job market. This may involve selling matches or shoelaces on the street, ice-cream vending, shoe-shining, rubbish collecting or scavenging bottles and other types of waste for recycling. Begging, petty crime and prostitution are other, less legal ways of scratching a living. These activities are part of the **informal economy**. They are 'informal' in that they operate in an essentially spontaneous and irregular way. They play an important part in the day-to-day survival of many people in cities across the world, particularly in LEDCs.

Another informal activity is paratransit. Minibuses, hand-drawn and motorised rickshaws, scooters and pedicabs (tricycles used as taxis) are the main paratransit modes. This informal transport system arises from a lack of other means of transport. Business is fairly good, but unfortunately these small-scale vehicles add to the problems of congestion on already busy roads (*Case study 27*).

Within the informal economy, it is possible to recognise distinct groups of workers (Figure 40). In Bangladesh, as elsewhere, there are significant differences between these groups when it comes to wages or earnings. Employers earn most, homeworkers least. There are also noticeable gender differences between the groups. The employers are mostly male, while the homeworkers are mostly female. In the other groups, there is a balance between men and women.

In Dhaka (Bangladesh) it is estimated that there are nearly half a million children involved in informal activities. Most of them work from dawn until dusk, earning on average 14 taka (about 12p) a day to help support their families. Their jobs vary from begging and scavenging (rag-picking) to domestic servants and money collectors on various forms of paratransit. These children work in vulnerable conditions. They are exposed to hazards, such as street crime, violence, drugs, sexual abuse, toxic fumes and waste products. These extremely poor working conditions lead to serious health and developmental problems.

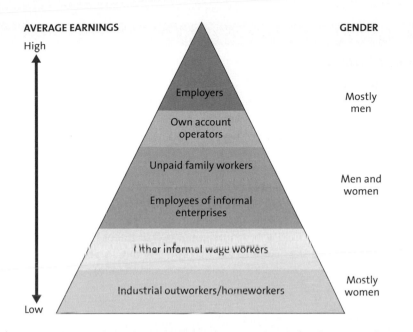

AVERAGE EARNINGS

High

GENDER

Employers — Mostly men

Own account operators

Unpaid family workers — Men and women

Employees of informal enterprises

Other informal wage workers

Industrial outworkers/homeworkers — Mostly women

Low

Figure 40 Gender differences within the informal economy

The benefits of the informal economy are being realised slowly. It provides a wide range of cheap goods and services that would otherwise be too expensive for many people. It also allows average wages to be kept low. This means that TNCs do not have to look elsewhere for cheap labour. Despite these benefits, one of the effects of informal employment is increasing poverty, particularly in urban areas. Informal jobs keep the city going, but they do not break the cycle of poverty. Other problems associated with the informal economy include:

- insecurity of work and income
- exposure to work-related risks
- no health, disability, unemployment or life insurance
- uncertain legal status
- few rights or benefits
- lack of organisation

Addis Ababa in Ethiopia is one of a growing number of cities beginning to see the significance of the informal economy and the need to encourage it. The city authorities are giving small loans and business services to help turn small, 'spontaneous' activities into self-sustaining, larger enterprises. These businesses could employ the city's pool of unemployed and underemployed people.

A flourishing informal sector can be a vital part of an LEDC city's economy, which was also true not so long ago of many MEDC cities. However, the introduction of laws and tough regulations designed to protect MEDC workers has effectively shut down most of the informal economy.

Of all the economic activities, both formal and informal, retailing may be the most readily associated with towns and cities. Much retailing is concentrated in the city's hub — the CBD — which enhances its perceived status in the portfolio of urban functions. Retailing is a dynamic part of the city's economy. It is particularly vulnerable to short-term shifts in the volume and tastes of consumer demand. In recent decades, it has experienced many changes relating to location and character.

Issues of change

Shifts in location

Over the last 50 years, the mobility of shoppers has improved dramatically due to increased car ownership. Less reliance on public transport systems in the CBD has allowed speculators to invest in shopping outlets in new locations (Figure 41). The direction of movement has been mainly outwards, towards the urban fringe and beyond, as shown by the huge retailing complexes (out-of-town shopping centres and hypermarkets) that have mushroomed on greenfield sites. Some of these developments have been located in order for them to be accessible from a number of towns and cities. For example, within a 45-minute drive of the huge retailing complex of Meadowhall, on the outskirts of Sheffield, is a population of over 2 million. The Bluewater shopping centre, located on the M25 close to the Dartford crossing, is able to attract customers from a large urban population living on both sides of the River Thames.

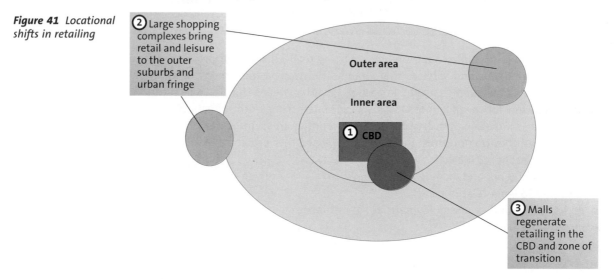

Figure 41 Locational shifts in retailing

② Large shopping complexes bring retail and leisure to the outer suburbs and urban fringe

Outer area

Inner area

① CBD

③ Malls regenerate retailing in the CBD and zone of transition

While major retailers have moved out from the CBD, shopping has become linked with leisure time — it is now seen by many as a recreational activity. Sunday opening has encouraged this consumer perception. This locational shift in retailing has hit the traditional CBD and created a 'doughnut effect' in city structure. Planners have now come to realise this serious downside to the decentralisation of retailing. Since 1994, the UK government has limited the number of new out-of-town developments. The CBD has had to fight back in these 'store wars' by improving:

- its image as a shopping centre
- the quality of the shopping environment and experience
- its accessibility

Park-and-ride schemes, pedestrianisation and the building of enclosed malls have been widely used to lure retailers and shoppers back into the CBD. Southampton has been one of the most successful in this respect, with its West Quay and other city-centre regeneration projects (*Case study 19*). However, many city centres are dealing with another potentially damaging issue — urban terrorism (*Case study 23*).

Shifts in character

'Westernised' consumer society is spreading rapidly to cities in the 'McWorld'. Fast-food chain McDonald's now has over 30 000 restaurants in 119 countries, which together serve 46 million customers every day. On its opening day in Kuwait City, the queue for the McDonald's drive-through was over 10 km long.

The concept of mall shopping can be seen from Addis Ababa to Putrajaya (*Case study 33*). In Putrajaya, retailing giants like Carrefour and Tesco are muscling in on the new hypermarkets. When it comes to retailing, everywhere is looking increasingly uniform.

This uniformity (**cloning**) is well demonstrated in the UK, as high streets lose their local identity. Independent butchers, greengrocers, pet shops and dry cleaners are being driven out by national supermarket retailers, fast-food chains, mobile phone shops and global fashion outlets. At present, property developers do not have to guarantee the provision of affordable premises for locally owned stores. On the other hand, market research surveys of UK shoppers reveal increasing dissatisfaction, particularly with the cloned architecture and retail opportunities of out-of-town centres. This dissatisfaction is encouraging for central city retailing, as there is a chance to capitalise on the unique historical character and layout of most city centres. Interestingly, perhaps the greatest retailing diversity in British cities today is in multi-ethnic areas, where variety is most evident in the range of food shops.

This case study is about important changes in retailing in the UK. However, these changes are global and not unique to this country. In the UK, the locational shift of retailing has been an important issue in the broader greenfield/brownfield debate.

Socio-cultural issues

Case studies 23, 24 and 25 are about disaffection and behaviour. In Case study 23, two kinds of urban terrorism are illustrated that particularly threaten central-city areas. Both are driven by alienation from modern society, but at different spatial scales. Case study 24 examines the disaffection rooted in the vicious cycle of poverty that condemns people to live as second-class citizens in many MEDC cities. Discrimination and exclusion based on ethnicity also give rise to disaffection and, as the case study shows, this is not always confined to the inner city. Case study 25 looks at an issue facing most MEDCs. Although a national concern, the problems of an ageing population are most keenly felt in the city.

URBAN TERRORISM

Case study **23**

The defence of urban space and society

International terrorism

Castles and defensive walls were a characteristic of many early British cities. Fortress architecture and defensible space have recently re-emerged as features of urban design, and we have become used to CCTV cameras, locked gates, barricades and bollards.

During the troubles in Northern Ireland, Belfast became famous for its ring of steel surrounding the main shopping area. Since the early 1990s, terrorists have targeted global cities in order to maximise media coverage and to hit at the heart of powerful economies. The 1992 and 1996 Canary Wharf bombs led to a more proactive response by urban decision-makers. The City of London (the famous Square Mile), which generates much of the UK's wealth, has effectively been turned back into a medieval fortress by its own ring of steel. The 30 entrances have been reduced to seven, and road checks and road blocks can be operated. Newer technology has meant CCTV coverage and automatic number plate recording can be linked instantaneously to databases. An interesting side-effect is the improvement in the overall environment as traffic has been reduced. The whole of London Docklands is now protected by a similar, well-advertised 'iron collar'.

Figure 42 The terrorist attacks on the World Trade Center, New York

TopFoto

Media coverage of the horrific 9/11 attacks on New York and Washington DC in 2001 brought to a global audience the potentially devastating impacts of terrorism (Figure 42). Governments and insurers alike were forced to take a new look at security in cities. London, because of its global city status, has been in the limelight for counter-terrorism. Buildings like the Houses of Parliament and the US Embassy have their own rings of concrete, camouflaged to make them more aesthetically acceptable. 'Terror architecture' is on the increase — new buildings are designed to reduce the lethal impact of terrorism. Urban zones are being mapped to show level of risk, based on function and building.

London's congestion charging system (*Case study 27*) has aided the 'landscape of surveillance' policy — it is estimated that the average Londoner is recorded on CCTV camera 300 times a day.

Some analysts advocate a decentralisation of key functions to reduce the risk of one-hit terrorism, such as at the twin towers of the World Trade Center in New York. So far, London has resisted this and is continuing with its high-rise and centralisation of core functions policies. However, the bombings of July 2005 have raised the question as to whether any measure can really counteract the threat of determined terrorists. There are also fears that rings of security may develop into rings of exclusion that impinge on free societies (*Case study 16*).

Central-city 'terrorism'

City-centre areas are not only threatened by international terrorism. In recent decades, many MEDC authorities have encouraged more people to live in town and city centres. They have done this to overcome the doughnut effect, by promoting 'living over the shop' schemes, converting derelict buildings and redeveloping brownfield sites for housing. At the same time, urban authorities have made attempts to promote their central areas as night-time 'playgrounds', by increasing leisure facilities, such as bars, nightclubs, cinemas and casinos. The aim has been to create a 24-hour city centre that strengthens the local economy and creates jobs. Sadly, what seemed to be a good idea at the time — bringing new life back into the 'dead heart' of the city — is now looking like a recipe for disaster. Residents in the city centre are having their lives 'terrorised' by antisocial behaviour — violence, noise and mindless vandalism. Other irritants include litter and youths vomiting and urinating in the streets. City and town centres, as well as some suburban neighbourhoods, are suffering from the spread of this unacceptable yob culture.

Research has shown a distinct correlation between this antisocial, often violent, behaviour and proximity to nightclubs and bars. Young males are the main perpetrators and victims of this violence, but young women are becoming increasingly involved. Many injuries occur in or just outside pubs and clubs. Violence is also associated with queues for taxis and fast food. The peak periods for violence are at weekends, when groups of yobs are frequently reported rampaging through city centres, such as in Nottingham and Plymouth. The principal cause of this yobbish behaviour is binge-drinking, often by under-age youths. This is encouraged by 'happy-hour' prices, unlimited licensing hours and lax enforcement of the law.

A wide range of short-term actions can be taken to reduce this problem. These include:
- imposing fixed, on-the-spot fines for drunk and disorderly behaviour
- inflicting stiffer penalties — fines, curfews, antisocial behaviour orders (ASBOs) and community service — for second-time offenders
- making parents responsible for the antisocial behaviour of their children
- introducing higher-profile policing

In the longer term, urban design can play an important part. For example, new housing in the Temple Bar area of central Dublin is designed to incorporate active, but quiet, uses, such as hairdressers on the ground floor with flats above. The flats are protected from the activities on the street by being arranged in a perimeter block around a quiet internal courtyard. Access to the courtyard and flats is limited to one gate, which is securely locked and under the control of the residents. Another strategy is to encourage entertainment venues away from residential quarters, as in the Riverside area of Norwich and Sheffield's Music Street. In Manchester, a surveillance ring involving over 400 CCTV cameras has been set up around the rejuvenated city centre following the 1996 bombing. The aim is to improve safety, reduce fear and crime, and to encourage the influx of shoppers, residents and businesses. Manchester is not alone in overtly linking security to regeneration.

Terrorism has become both a global and a local urban issue. Mindless behaviour, whether by misguided zealots or unruly youths, makes a mockery of the once widely held belief that cities and civilisation are synonymous. The history of urban design is full of examples of the built-up area responding to changing circumstances. These two brands of terrorism are simply triggering a set of responses that hopefully will provide the required level of security and urban resilience.

TENSIONS IN THE SUBURBS

Case study 24

France today, where next?

During late October and November 2005, a period of sustained unrest took place across France. Looting, arson and street rioting erupted, initially in the suburbs of northeast Paris, but soon spreading elsewhere in the capital (Figure 43) and to other French cities, stretching from Lille to Toulouse.

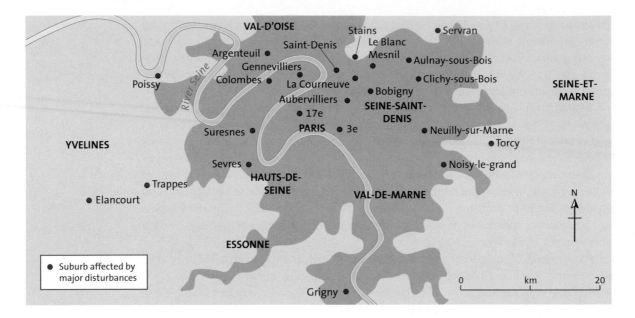

Figure 43 *Riot-torn suburbs in Paris, October–November 2005*

The riots were initiated by the death of two North African youths, accidentally electrocuted when they entered an electricity substation. The claim that these youths were being harassed by white police soon gave vent to a simmering and deep-seated resentment among the largely immigrant population.

Immigrant populations here and elsewhere in France believe they are being treated by the white French as less than second-class citizens. They are forced to live in areas of substandard housing. Unemployment rates among immigrant males are almost 50% in some suburbs. The tower block estates, often built without supporting services and employment opportunities, have become **sink estates** for the poorest workers moving into France, particularly from Africa. Many of these people are the victims of social exclusion.

Only a small proportion of people of North African origin — who are referred to as 'beurs' — manage to work their way out of these estates of despair. One of them, Aziz Senni, has recently written a book about his experiences and views, the title of which translates as *The social lift has broken. I took the stairs*. This contains a number of important messages from a moderate Muslim proud to be French:

- Job creation should be a top priority, along with the provision of better schools and educational opportunities.
- The racial discrimination that leads to immigrant workers being the first to be made redundant must be outlawed. The factor that 'broke the lift' was the era of high unemployment that started in the late 1970s. Senni claims that this stranded the ethnic Arab and black estate dwellers in the 'basement'.
- The government needs to do more to rectify its neglect of impoverished suburban estates.
- Ethnic minorities must use their own resources to climb out of the ghetto and avoid the well-trodden route of petty crime.
- Most worrying is Senni's view that the next revolt on the French housing estates 'will be more explosive; they will use military weapons. They already have Kalashnikovs and rocket launchers in there.'

Contemporary Case Studies

Mindful of what has happened in France, is there any chance of this happening in the UK? We would be ignoring reality not to recognise that there are two potential hot spots:

■ the poorest social housing estates
■ those areas where ethnic minorities are most concentrated

Over recent decades, the worst housing estates have acquired a bad name. As this has happened, those residents alarmed by the declining reputation, and with the ability to move out, have done so. They have left behind an increasing concentration of the poorest and most disadvantaged families. This has led to addressism. Residents of these now stigmatised estates find it difficult to get a job because employers discriminate against them on the basis of their address. Figure 44 shows the downward spiral leading to the sink estate stigma.

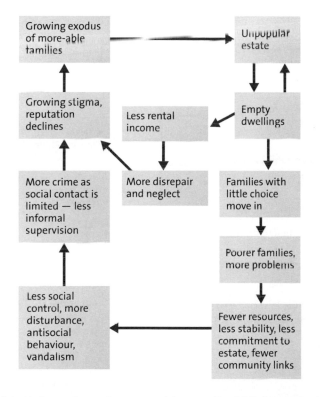

Figure 44 The downward spiral of a sink estate

Although 'ghetto' may be a strong word to use, the UK's increasingly multi-ethnic society shows high levels of segregation. Figure 45 suggests that the concentration, for example of south Asian or black Caribbean communities in some inner-city areas, be it of London or Bradford, may not be simply a matter of being forced into areas of poor housing. There may be an element of choice. In this new age of urban terrorism and widespread fear (*Case study 23*), minority concentrations may easily, and wrongly, be looked on with suspicion as harbouring, say, some dissident group.

In both sink estates and ghettos, the question of what best to do raises a whole new set of debatable issues, such as redevelopment versus improvement and concentration versus dispersal/assimilation.

This case study is a salutary warning about the dangers of neglecting the issue of social exclusion. France is not the only country at risk. The UK could be the next in line.

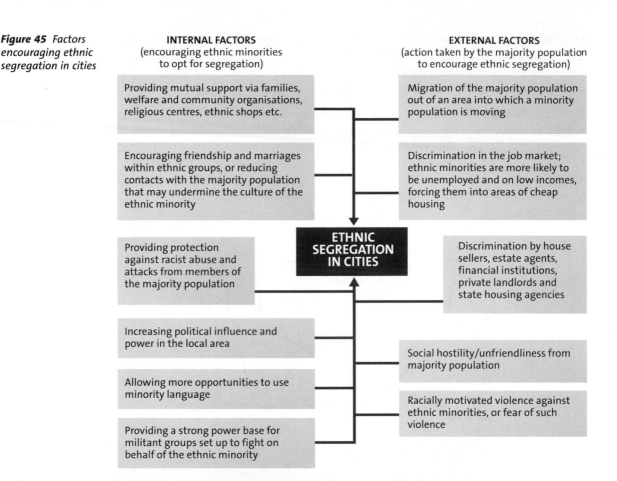

Figure 45 *Factors encouraging ethnic segregation in cities*

Providing mutual support via families, welfare and community organisations, religious centres, ethnic shops etc.

Migration of the majority population out of an area into which a minority population is moving

Encouraging friendship and marriages within ethnic groups, or reducing contacts with the majority population that may undermine the culture of the ethnic minority

Discrimination in the job market; ethnic minorities are more likely to be unemployed and on low incomes, forcing them into areas of cheap housing

ETHNIC SEGREGATION IN CITIES

Providing protection against racist abuse and attacks from members of the majority population

Discrimination by house sellers, estate agents, financial institutions, private landlords and state housing agencies

Increasing political influence and power in the local area

Allowing more opportunities to use minority language

Social hostility/unfriendliness from majority population

Providing a strong power base for militant groups set up to fight on behalf of the ethnic minority

Racially motivated violence against ethnic minorities, or fear of such violence

19 Question

Using case studies

Copy Figure 44 and annotate it to show how the downward spiral can best be tackled by managers to reduce the 'sink' elements of poverty and antisocial behaviour.

Guidance

Include the following measures: economic (retraining and accessibility to work), social (education and community policing), and environmental (renovation and landscaping).

Case study 25 CITIES AND THE THIRD AGE

Elderly issues

The UK is one of an increasing number of MEDCs with a significantly ageing population. By 2020, more than half of the nation's adults will be over 50, and most of these will live in an urban environment.

The issues shown in Figure 46 need planned responses. Although the economic ones may be the most challenging, the more serious among them (pensions, the dwindling labour force etc.) are for national, rather than city, decision-makers to deal with. It is the social and environmental issues that city authorities are expected to resolve.

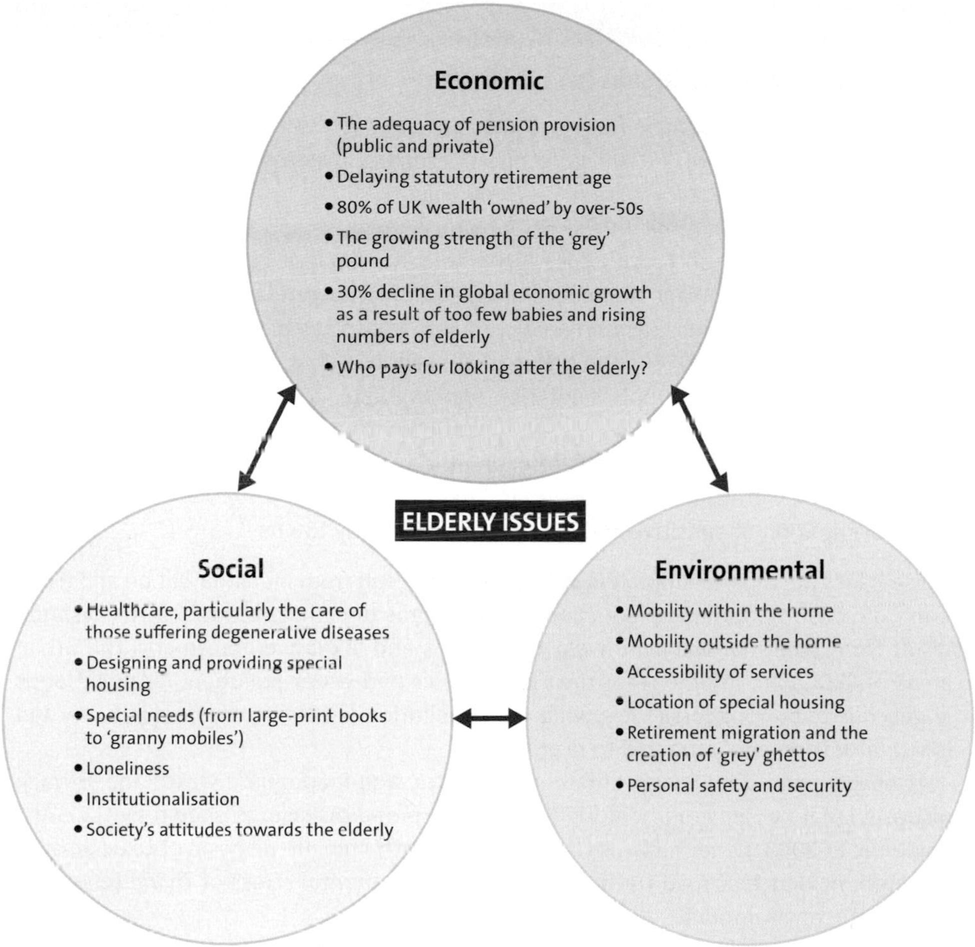

Economic
- The adequacy of pension provision (public and private)
- Delaying statutory retirement age
- 80% of UK wealth 'owned' by over-50s
- The growing strength of the 'grey' pound
- 30% decline in global economic growth as a result of too few babies and rising numbers of elderly
- Who pays for looking after the elderly?

ELDERLY ISSUES

Social
- Healthcare, particularly the care of those suffering degenerative diseases
- Designing and providing special housing
- Special needs (from large-print books to 'granny mobiles')
- Loneliness
- Institutionalisation
- Society's attitudes towards the elderly

Environmental
- Mobility within the home
- Mobility outside the home
- Accessibility of services
- Location of special housing
- Retirement migration and the creation of 'grey' ghettos
- Personal safety and security

***Figure 46** Elderly issues in the city*

The solutions are not simple, as the issues are interlinked. To meet those social and environmental needs requires funding. A city with more than half of its population not working is hardly likely to have the necessary financial resources.

There is an interesting political twist to this scenario. A growing number of elderly people in the electorate means a strengthening of the 'grey' lobby. How long will it take our political parties to wake up to the significance of the growing 'grey' vote? The political situation could easily be made to suit the elderly. It needs only a gentle mobilisation of the ranks of the elderly to insist that their special needs are met.

It is not just the MEDCs that have ageing populations. Are the elderly issues any different in LEDCs? Are LEDCs coping any better than MEDCs?

Environmental issues

Almost all the case studies considered so far have an environmental dimension. In the next three, this dimension is even more prominent. *Case study 26* focuses on LEDCs and what needs to be done to improve the built environment.

Case study 26 THE BROWN AGENDA

Environmental quality in an LEDC city

'Good environmental conditions are not a luxury and cannot wait — they are an important contributor to quality of life and to economic competitiveness.' (Kristalina Georgieva, director of the World Bank Environment Department)

The 1992 Earth Summit in Rio de Janeiro, and most subsequent literature and debates, have focused on 'green' issues such as biodiversity loss, global warming, climate change, ozone depletion and marine pollution. The **Brown Agenda**, launched at the same summit, focuses on the environmental problems associated with the development process. The major issues raised in the agenda include:

- the lack of safe water supply, sanitation and drainage
- inadequate solid and hazardous waste management
- uncontrolled emissions into the atmosphere
- accidents — traffic, industrial, environmental
- the occupation of sensitive and unsafe land by shanty towns

These issues are most keenly felt in LEDC cities and result from industrialisation and rising levels of consumption and production. The problems directly linked to industrialisation result from poor management, weak regulations and lax law enforcement. The urban areas of LEDCs are infamous for their levels of air and water pollution. Air pollution is largely due to burning fossil fuels, while water pollution results mostly from industry and informal settlements attracted to river banks.

Densely packed housing, contaminated water, and inadequate waste and sewage disposal create an environment highly likely to spread disease, as shown by the SARS epidemic in 2003. Other hazards to health include the unsafe disposal of toxic wastes, the high incidence of road traffic accidents and the mental stress of trying to survive in a hostile environment.

The following examples cover two Brown Agenda issues: solid waste (Dhaka) and housing the poor (Addis Ababa). Water pollution, with reference to Shanghai, is examined in *Case study 29*.

Solid waste

Dhaka, a fast-growing mega-city, generates about 3000 tonnes of solid waste per day. Only 42% of this enormous quantity of waste is collected. The rest rots on roadsides, in open drains and on low-lying areas. It is a major contributor to the deteriorating quality of the city's environment. Waste Concern, a research-based non-governmental organisation (NGO), initiated a pilot project in the 1990s to investigate ways of recovering value from solid waste. They came up with a community-based collecting and composting scheme.

The scheme is low cost and relies on simple technology. The compost site can be located near the residential areas it serves, as the nuisance caused by odours and flies is minimal. It produces an environmentally safe product that finds a ready market in the agricultural areas adjoining the city. The money earned provides wages for those employed in door-to-door collection of the rubbish (for which there is a small charge). The organic waste is used to make compost. Every day, 500 kg of compost is produced from 2 tonnes of solid waste. At present, the non-organic waste is moved to designated landfill sites, but there is scope for sorting and recycling, perhaps, as in China, using some of it to manufacture bricks. The scheme:

- helps to create a cleaner environment
- creates much-needed jobs for the poor
- reduces the rate at which landfill sites are used
- makes a cheap fertiliser

The hope is that such schemes will catch on and spread throughout the city.

Shanty towns and slums

The most keenly felt environmental issue in many LEDC cities is housing — especially for those trapped in the cycle of poverty. Adequate accommodation is a basic human need. In cities such as Addis Ababa, Dhaka and Mexico City, the cheapest housing is usually provided in shanty towns or squatter settlements. Shanty towns are usually located in areas where land values have been depressed by their risk of environmental disasters, such as flooding, landslides and pollution (Figure 47). These slums are built on areas avoided by formal housing, and they are seen by many as not only inevitable but also a mark of a city's success. They will continue to appear and persist as long as people are excluded from affordable housing by low wages. Sadly, minimal wages almost guarantee a city's economic success. Where is the incentive to raise the wages of the poor?

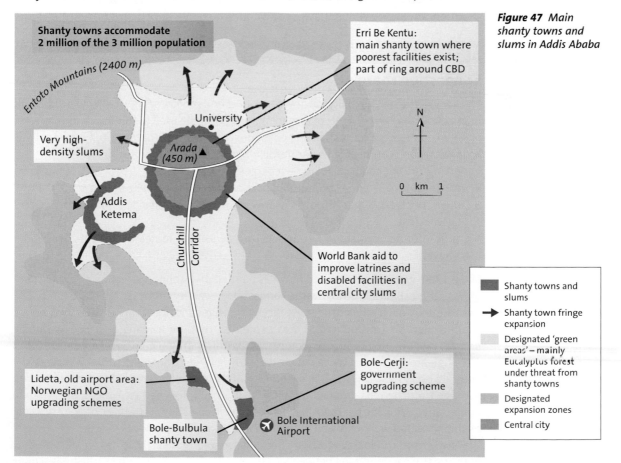

Figure 47 *Main shanty towns and slums in Addis Ababa*

These settlements of slum housing scar the cityscape. The challenge is what to do with them. There is a range of options:

- raze them to the ground and let the poor perish — a solution favoured by President Mugabe in Zimbabwe
- redevelop the sites completely and build cheap, high-rise housing for rent — not easy where most people live a hand-to-mouth existence rooted in the informal economy, such as in Dhaka (*Case study 21*)
- leave them as they are and hope that, in time, residents will have the means to upgrade their homes — the most widely adopted option to date
- introduce site-and-service schemes that provide the basic amenities of water supply, sewage disposal and electricity
- introduce self-help schemes that provide grants, cheap loans or even the materials for residents to replace their shacks with properly constructed dwelling units

The last two actions are known as **consolidation**.

The Brown Agenda is broadly about environmental improvement. That improvement should extend to include both the living conditions and the health of the poor.

Case study 27 illustrates the point that many city issues are multi-faceted, simply due to the links or inter-connections that exist. Actions taken to deal with a particular situation almost inevitably have positive and negative knock-on effects elsewhere.

Case study 27 | TRAFFIC MANAGEMENT

Congestion charging

The long-term **sustainability** of any town or city depends on how effective its traffic management system is. A mix of technology and policy are used to combat traffic problems of speed, flow, safety, congestion and pollution. The main types of traffic management are:

- segregation — rerouting heavy traffic, ring roads, pedestrianisation
- parking pricing and availability — charges, permits, spaces
- vehicle control — odd/even number licence plates, traffic light systems
- public transport — bus priority systems, park-and-ride, subsidies, alternative modes (e.g. metros)
- traffic information systems — advising travellers about routes, black spots and modes of travel

Since 1924, when the first white line was painted on a London street as an experiment to ease the capital's growing traffic congestion, a variety of measures has been put to the test. In London, one of the most recent is congestion charging. In 2003, just before charges were first introduced, it was estimated that, on average, just over 1 million people were entering central London every weekday morning between 7a.m. and 10a.m. Although around 80% of those people were using public transport, 40000 vehicles were entering every hour. Drivers were spending half their travel time at a standstill in queues. Average traffic speeds dropped to below 10mph and, for the first time, there were no longer any rush hour peaks — just 12 hours of congestion. Estimates of the true cost of this situation were between £2 million and £4 million per week.

The central London congestion charging scheme requires that nearly all motorists entering the charging zone between 7a.m. and 6.30p.m. on weekdays must pay a daily fee of £8. The only other British city to attempt such a scheme is Durham, where traffic in the charging zone has shrunk from 2000 to 200 cars per day. Overseas, Singapore has been a leading light. In 1975, it introduced a licensing scheme that charged motorists entering the downtown area. This has now been superseded by a state-of-the-art electronic road-pricing system. This requires that all vehicles wishing to enter the restricted zone must be fitted with an on-board unit. Charges are automatically deducted from a cash card slotted into that unit. Gantry cameras at the entrance to the zone record the number plates of vehicles without the required unit and drivers are fined heavily. Some cities in the Netherlands and Sweden have similar schemes in place.

There has been great interest in the London experiment. Other cities, including Manchester and Hong Kong, have been watching and weighing up the pros and cons (Table 4). London generally regards the scheme as successful, and in February 2007 the charging zone will be extended westwards (Figure 48).

Table 4 The advantages and disadvantages of congestion charging

Advantages	Disadvantages
■ Congestion levels lowered (by 30% in central London)	■ Reduction in the number of people coming into the city centre for reasons other than work — shopping, entertainment etc.
■ Fewer vehicles on the road (16% drop in central London)	■ Loss of commercial revenue as a result
■ Journey times reduced (by 14% in central London)	■ Traffic displacement around charging zone making situation worse elsewhere
■ Reduction in the number of road accidents (by 20% in central London)	■ Hits less well-off people who are car-dependent
■ Greater use of other modes of transport such as bicycles and scooters, and more car sharing	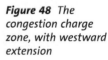
■ More business for public transport	
■ Better bus and taxi services	
■ More efficient distribution of goods and services	
■ Lowering of pollution levels — exhaust emissions and noise	

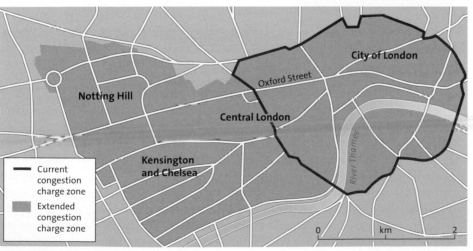

Figure 48 The congestion charge zone, with westward extension

While congestion charging delivers environmental benefits, it can also lead to both costs and benefits in economic and social terms.

Congestion charging is only one of a number of strategies being adopted to ease the traffic and transport problems of cities. Others are mainly linked to public transport and trying to make this a more attractive (i.e. cheaper and quicker) option than the car.

Case study 28 focuses on risk, which has always been an issue for cities. In early times, people may have been more aware of the risks posed by natural hazards. Was the city safely located with respect to known natural events such as floods, volcanic eruptions, earthquakes and tropical storms? As shown in Figure 49, traditional risks are made up of more than just natural hazards — there are others associated with the natural environment. A new generation of risks has since emerged. These involve those brought about by humans, either through their abuse of the natural environment or by the nature of the built environment. Traditional risks, however, have not gone away — in particular, the threat of natural hazards remains. Natural hazards still challenge the technology and resources of even the most affluent cities. New Orleans is one of the latest examples.

Figure 49 *Changing environmental risks and city income levels*

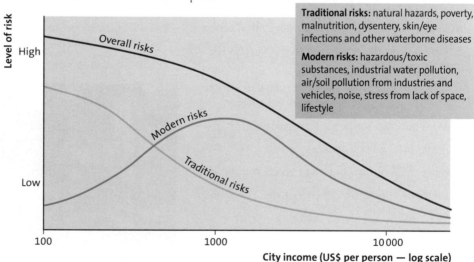

Traditional risks: natural hazards, poverty, malnutrition, dysentery, skin/eye infections and other waterborne diseases

Modern risks: hazardous/toxic substances, industrial water pollution, air/soil pollution from industries and vehicles, noise, stress from lack of space, lifestyle

LIVING WITH RISK

New Orleans will never be the same again

New Orleans has learned to live with risk. This is true of any city unwisely located on the delta of one of the world's rivers most renowned for its tendency to flood, as well as in the pathway taken regularly by immensely powerful hurricanes.

To become home to around 1.3 million people, New Orleans has had to survive a long history of natural disasters. When Hurricane Katrina swept overhead in the autumn of 2005, the city experienced a natural event that crossed the threshold between calamity and total disaster. The hurricane created a lethal cocktail of viciously high winds, torrential rain and a major storm surge that backed up the waters of the Mississippi River and caused flooding to record levels (a 1 in 500-year event). When the levées protecting New Orleans gave way, 224 billion gallons of water spilled from Lake

Pontchartrain (a flood relief lake) and flooded 80% of the city. An area the size of the UK was inundated and over 1000 people killed.

This happened in the world's 'most powerful' nation. Although warning systems were in place, flood and hurricane protection measures were ineffective, particularly when the badly constructed levées were breached. The emergency and rescue operations to help the stricken city in the days immediately following the hurricane were also sadly lacking.

So great was the damage to New Orleans that, in the immediate aftermath of the hurricane, serious consideration was given to abandoning the city altogether. However, damage to the vital river port and the historic French quarter — the focus of the city's tourism — was found to have been much less than originally feared, and large areas of the suburbs were not flooded at all. It was decided to start rebuilding the extensively damaged parts of the city; a task that will take at least 2 years.

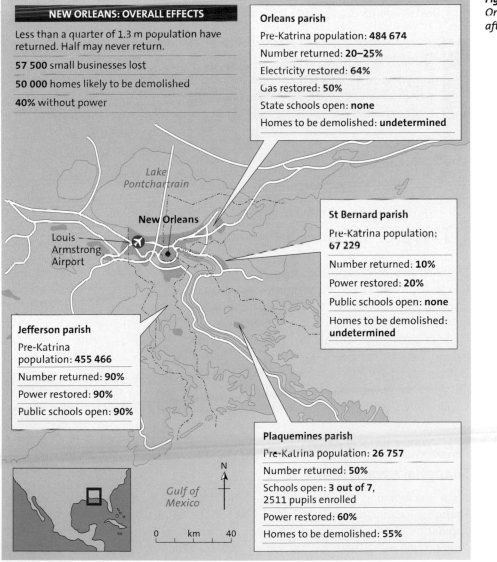

Figure 50 New Orleans 11 weeks after Katrina

NEW ORLEANS: OVERALL EFFECTS

Less than a quarter of 1.3 m population have returned. Half may never return.

57 500 small businesses lost

50 000 homes likely to be demolished

40% without power

Orleans parish

Pre-Katrina population: **484 674**

Number returned: **20–25%**

Electricity restored: **64%**

Gas restored: **50%**

State schools open: **none**

Homes to be demolished: **undetermined**

Lake Pontchartrain

New Orleans

Louis Armstrong Airport

St Bernard parish

Pre-Katrina population: **67 229**

Number returned: **10%**

Power restored: **20%**

Public schools open: **none**

Homes to be demolished: **undetermined**

Jefferson parish

Pre-Katrina population: **455 466**

Number returned: **90%**

Power restored: **90%**

Public schools open: **90%**

Plaquemines parish

Pre-Katrina population: **26 757**

Number returned: **50%**

Schools open: **3 out of 7,** 2511 pupils enrolled

Power restored: **60%**

Homes to be demolished: **55%**

Gulf of Mexico

N

0   km   40

One sensitive issue associated with Hurricane Katrina is that it inflicted most damage on low-lying residential areas, largely inhabited by the poor black population (the population of New Orleans is 65% black).

Figure 50 shows the situation in New Orleans 11 weeks after Katrina. The scale of the recovery task was then indicated by the following facts:

- It was expected to take 6 months to restore electricity to the worst-hit areas.
- The state of Louisiana already faced a $1 billion budget deficit.
- Thousands of families were left with mortgages on homes that no longer exist.
- Up to half of the city's 115 000 small businesses were thought to have been lost for good.
- Katrina created 38 million m³ of debris; the equivalent of five American football pitches each piled 1.6 km high.

The case of New Orleans is a sobering reminder that, even in the most developed parts of the world, there are times when nature has the upper hand. Deciding to rebuild New Orleans on its original site is a considerable gamble.

20 Question

Copy Figure 49 and add another curve to show how social risk (terrorism, personal safety, crime, vandalism and other forms of antisocial behaviour) might change with city income. Briefly justify what you have drawn.

Guidance

Read *Case studies 16, 21, 23* and *24* again.

Image issues

All cities that have ambition regarding their futures will be concerned with their **image**. That concern may be felt most keenly in those cities shifting their economies from industry to services. However, in this age of economic globalisation, the question of image is hardly any less significant in those cities and countries less advanced along the urban and development pathways. A city wishing to attract inward investment needs to 'look good'. Many of the issues considered in this chapter can easily tarnish that much-needed image. *Case study 19* has already provided some information about this topic. *Case study 29* examines a different development scenario. It also allows links back to *Case study 12*.

Case study 29 | PUTTING A NEW FACE ON IT

Shanghai's waterfront

Since the opening of China and the transition to a socialist market economy, Shanghai has mushroomed to become one of the largest cities in the world with a population of 14 million people. The city is now estimated to be the world's largest construction site. There are currently 4000 buildings over 24 storeys high, with another 1700 under construction or with planning approval. The focus of this activity at the moment is the Pudong area on the west side of the Huangpu River and adjacent to the Old City (Figure 29). The prestigious, high-rise buildings appearing along or close to the waterfront are a vital part of Shanghai's re-imaging as a global city.

The Shanghai waterfront consists not only of the Huangpu River, but also Suzhou Creek. This used to accommodate most of the city's port operations, warehousing and industry. The main port functions and much of the industry have now been relocated on the Yangtze River. This has created an opportunity to redevelop the waterfront. One of the main aims of the current scheme is to make the waterfront an asset for all by creating a linear public open space, backed by the new generation of prestige buildings (Figure 51).

The challenge in Shanghai has not only been to transform the waterfront, but also to reduce pollution of the rivers. The Suzhou and Huangpu rivers are the main sources of drinking water and also the sewers. According to the World Bank, 4 million urban residents live in crowded conditions, with inadequate access to drinking water and sanitation. Less than 60% of waste water and storm water in the city is hygienically intercepted and disposed of. This is hardly the image city authorities are trying to project. They hope that the longer-term spin-off of the image-improvement exercise focusing on the waterfront will be:

- a booming city
- more funds to pay for these infrastructural improvements
- a real prospect that all households will be properly serviced in terms of piped water, electricity supply and waste disposal

Figure 51
The Shanghai waterfront

The image of a city has many different facets. Image is not just about what the city looks like but also its reputation in a wide range of features, from its economy to its human resources, and from its stability to good governance.

Government issues

Underpinning all the issues considered in this part is government direction. It should be understood that governments are involved to varying degrees in helping to deal with these sorts of issues. Intervention is more frequent in socialist economies, but these economies have dwindled in number. In market economies, the level of intervention varies from high (in the case of Singapore) to low (in many LEDCs).

Even within individual countries, the level of intervention may vary over time. In the UK during the 1960s and 1970s, there was a general consensus that the problems of our cities were best dealt with by state planning and state spending. During the 1980s and 1990s, mainly under Conservative rule, state support was gradually removed. The role of government was simply to initiate schemes (such as the Urban Development Corporations) and standard measures (such as deprivation indices), and then to leave cities, private businesses and groups of individuals free to act as they thought best. Ironically, this private-sector philosophy is being perpetuated now by a 'socialist' government.

Question

Investigate deprivation in a town or city near you.

Guidance

According to the UK government, deprivation is the lack of the type of diet, housing, work, environment, education, activities and facilities that are thought to be the basic norms for all citizens. Deprivation is therefore a relative and multifaceted concept.

If anything is to be done to remedy deprivation, then first we must measure it as accurately as possible and find out where it occurs (i.e. map it). In 2004, a new multiple deprivation index (MDI) was drawn up based on the 2001 Census data. Raw data relating to seven key quality of life indicators, called domains, are collected for a scheme of small spatial units known as super output areas (SOAs). The weighting given to these seven domains is shown in the following table.

Domain	Weighting (%)
Income deprivation	22.5
Employment deprivation	22.5
Health deprivation and disability	13.5
Education, skills and training deprivation	13.5
Barriers to housing and services	9.3
Crime	9.3
Living environment deprivation	9.3

Having calculated the MDI for each SOA, it is possible to produce a choropleth map to show the general pattern of deprivation, and more particularly to reveal the most deprived areas.

Each of 32 482 SOAs and 354 local authorities in England has been scored. The local authorities have also been ranked according to their overall MDI value. For example, Bradford and Southampton are ranked 31 and 96 respectively of the most deprived cities.

Log into one of the following websites and download MDI values for the SOAs of a town or city near you: www.direct.gov.uk/Dl1/Directories/LocalCouncils/fs/en
www.odpm.gov.uk/index.asp?id=1128440
www.scotland.gov.uk/stats/simd2004/
www.lgdu-wales.gov.uk/eng/WimdProject.asp?id=2077

Plot these values and identify the most deprived areas. Describe the overall pattern and the physical characteristics of the most deprived areas.

Making cities more sustainable

Sustainability is one of today's 'buzz' words and it is used rather loosely. In general terms, a sustainable city is one that:
- meets the needs of its present and future inhabitants
- has a minimal ecological footprint

This concept may be best understood if the city is thought of as an open system, with inputs and outputs (Figure 52).

External actions	
Input actions	
• Conserve natural resources	
• Ensure efficient use of resources	
• Protect biodiversity	
• Respect environmental capacity	

Internal actions
- Recycle waste
- Create a fairer society: eliminate exclusion
- Encourage wide participation in decision making
- Make living space healthy and secure
- Provide a 'green' infrastructure
- Reuse brownfield sites
- Make cities more compact and reduce use of private cars

Output actions
- Minimise emissions and pollution
- Restrict use of greenfield sites
- Provide leisure and recreational opportunities

Remember:

inputs + outputs = ecological footprint

Figure 52
Making cities more sustainable

Inputs, processes and outputs

Both the inputs and outputs of a city threaten sustainability. For inputs, the problem is the consumption of resources, particularly non-renewables. Because of their affluence, cities can consume huge quantities of resources and threaten to exceed the Earth's carrying capacity. However, outputs — such as waste, pollution and the spread of the built-up area — are even more of a threat to the environment.

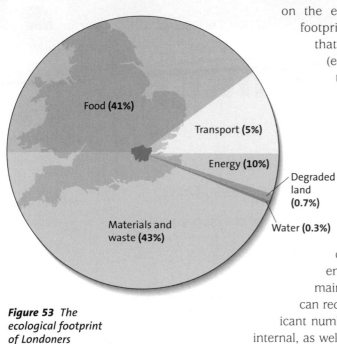

These combined impacts — inputs on resources and outputs on the environment — create the city's ecological footprint. A recent investigation of London revealed that each resident required 6.63 global hectares (equivalent to 8 football pitches) to provide for their current levels of consumption and to absorb their wastes (Figure 53). The current calculation for a sustainable footprint is 2.18 global hectares. In other words, Londoners consume more than three times their fair share of the Earth's resources.

However, a sustainable city is not just one with a light ecological footprint. Within the city system itself, 'costs' — such as congestion, stress, poverty, discrimination and crime — result from its various inputs (capital, employment, etc.) and processes. These costs are mainly economic, social and environmental. They can reduce the aspirations and quality of life of significant numbers of city dwellers. City sustainability has an internal, as well as an external, dimension.

Figure 53 The ecological footprint of Londoners

Raising sustainability

Many statements have been made that attempt to summarise the spirit of a sustainable city. Statements that set out what needs to be done in order to raise a city's sustainability may be more useful. For example, the Cambridge Local Agenda 21 sets out five objectives:
- increasing social equity and creating a fairer society
- encouraging participation, so that everyone can have their say
- improving living spaces
- maintaining surroundings and health
- conserving natural resources

The wide range of issues involved in moving towards urban sustainability can be found in the *Gaia Atlas of Cities* (Gaia Books, 2005). These include:
- resource budgeting
- energy conservation and efficiency
- renewable energy technology
- long-lasting built structures
- proximity between home and work
- efficient public transport systems
- waste reduction and recycling
- organic waste composting
- circular metabolism
- supply of staple foods from local sources

Although this use of the word 'sustainable' is relatively new to our vocabulary, the basic idea behind it has a long history, particularly in terms of building and expanding cities. The focus in the past was more on quality of design and habitability. More recent city landscapes in the UK have been influenced by a number of reformers, planners and architects in search of a better urban future (*Case study 30*).

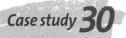

THE URBAN FUTURE

Case study **30**

Two visionaries

This case study looks briefly at the work of two people whose visions of the urban future have had, or continue to have, a considerable impact on the built-up area of British cities.

Ebenezer Howard (1850–1928)

During the nineteenth century, London became the largest city in the world. It was a place of both opportunity and squalor. Ebenezer Howard founded the Garden City Association, whose mission it was to solve the problems of the congested city and the 'undeveloped' countryside. This was to be achieved by combining the advantages or benefits of both town and country living in one place, and to avoid their downsides. The organisation's slogan was to build 'cities within gardens; gardens within cities'. The garden city model showed a central park surrounded by housing, with factories on the outskirts, all enclosed within a greenbelt. The aim was to create independent cities, rather than satellites of London.

The first garden city venture was at Letchworth in 1903, followed by Welwyn Garden City in 1920. These served as models for the New Towns built in the UK after the Second World War. While they had specific objectives, such as accommodating London overspill and revitalising declining regions, New Towns were also driven by notions of sustainability — expressed in terms of fusing the best aspects of urban and rural living. Another associated venture — Hampstead Garden Suburb, built in London in 1907 — had an influence on the character of much suburban development, particularly before the Second World War (1939–45).

Charles Edouard Le Corbusier (1887–1965)

Charles Edouard Le Corbusier was the most influential, admired and maligned architect of the twentieth century. Through his writing and buildings, he was the main player in the 'modernist' urban movement. His visions of homes and cities were as innovative as they were influential. Many of his ideas on urban living became the blueprint for postwar reconstruction. The many failures of his would-be imitators led to him being blamed for the problems of Western cities in the 1960s and 1970s.

Le Corbusier's early work involved designing a dwelling that could be mass produced, helping to solve the chronic housing problems of industrialised countries in the early twentieth century. His prototype was made of reinforced concrete. None of the internal walls was load-bearing, so the interior could be arranged as the occupant wished.

Despite his love of modern building techniques, Le Corbusier was determined that his architecture should improve nineteenth-century cities, which he saw as chaotic and 'dark prisons' for many inhabitants. Le Corbusier was convinced that a rationally planned city, using standardised housing types, could offer a healthy, humane alternative. His vision was of a decongested city with large blocks of flats towering over open, tree-dotted space.

Cities & Urbanisation

Figure 54
*Unité d'Habitation,
Marseilles*

After the Second World War ended in 1945, and with Europe's housing problems worse than ever, Le Corbusier got his chance to put his urban theories into practice. The Unité d'Habitation in Marseilles, built in 1952, is a product of three decades of Corbusian domestic architecture and urban thinking (Figure 54). Seventeen storeys high and designed to house 1600 people, the Unité incorporated various types of apartment, shops, clubs, recreational facilities and meeting rooms, all connected by raised 'streets'. It is now a popular building, and a coveted address for Marseille's middle-class professionals.

When Le Corbusier died in 1965, the backlash against urban modernism was gaining momentum. His theories on urban renewal (redevelopment rather than improvement) were plagiarised by UK local authorities on tight budgets. Slums were cleared, only to be replaced by 'human filing cabinets' — gaunt, poorly constructed, high-rise blocks. However, blaming Le Corbusier as the architect of postwar housing failure ignores his deep concern for human comfort and health.

Howard and Le Corbusier were two visionaries who have left their mark on Britain's towns and cities in the twentieth century. Twenty-first century visionaries might include Sir Peter Hall (see *Case study 36*, p. 98) and Richard Rogers. Rogers is renowned for his architectural designs, which incorporate the concept of sustainability. His more famous projects have been Canary Wharf (London), the Pompidou Centre (Paris), and Shanghai's CBD redevelopment. He has recently been commissioned to design Tower 3 on the site of the World Trade Center in New York.

Website: **www.richardrogers.co.uk**

Case study **31**  MILTON KEYNES

An early step in a sustainable direction?

Milton Keynes was designated, in 1967, as part of the British New Towns Programme. It was located midway between London and Birmingham and was designed to accommodate urban overspill, particularly from London, and eventually to play its part in creating a megalopolitan structure that would eventually incorporate the four major conurbations of Greater London, Birmingham, Merseyside and Manchester. Although always claimed to be a third-generation New Town, it was not a totally greenfield development — the chosen site already contained a number of towns and villages, with a total population of around 60000. The development of the site was planned to accommodate a population of 250000 by the early twenty-first century.

Milton Keynes has been a success in demographic and economic terms. It now has a population of around 215000. It has become a major employment node (over 115000 jobs), as well as a regional service centre, with a population of just under 8 million people living within an hour's drive. Three-quarters of the jobs are in the service sector (retailing, education, training, software and hardware design, banking, insurance and management consultancy).

Contemporary Case Studies

In terms of design and layout, Milton Keynes has also broken some new ground in the UK (Figure 55):

- All economic and social activities, such as jobs and leisure opportunities, are deliberately dispersed throughout the city and not segregated into special zones.
- Housing areas are deliberately 'mixed' (social and owner-occupied) to encourage social interaction.
- The original plan to build a new city based on a monorail public transport system was abandoned for a car-orientated city.
- There is a US-style grid road system, designed to even out traffic flows and avoid rush-hour congestion associated with typical radial town plans. Dual carriageway roads for through traffic intersect with local roads at approximately 1 km intervals.
- The city centre is a dense mixture of retail, leisure and offices. Its focus, a large central shopping area, is in the Guinness Book of Records for being the longest shopping mall in the world and the first covered shopping mall in the UK. It has 600 000 visitors a week.
- Nearly a quarter of Milton Keynes is parkland. A network of linear parks follows the course of the Grand Union Canal and the flood plains of the Rivers Ouse and Ouzel.

Figure 55
Milton Keynes strategic plan, 1989

Legend:
- Residential area
- Employment area
- Secondary and higher education
- Centre
- Services and community use
- Open spaces and recreation, including buildings
- Lake
- Reserve site
- Brickfield
- Railway
- City road
- Local road
- Motorway
- Boundary

0 km 3

The sustainability focus in Milton Keynes has been on the 'city system' itself (Figure 52). However, two significant shortfalls are apparent:

- The car-based nature of the city means that there is a heavy ecological footprint in terms of energy consumption. The strategy to disperse activities, and the overall low development density, mean that promoting public transport is not a viable option.
- With a housing stock of about 87 000 dwellings and prices rising rapidly, Milton Keynes is one of the most expensive areas in the UK outside London. Demands for affordable and rented homes have been neglected by developers. There is a growing degree of exclusion.

Looking at inputs, the ecological footprint is deepened not just by the high energy demands of private transport, but also by the fact that future expansion will consume 1800 hectares of greenfield land. Milton Keynes does not have a stock of brownfield sites.

All new housing will be built to high energy-efficiency standards and higher housing densities will help the provision of public transport and affordable homes. The expansion will involve a redesigned retail centre.

As for outputs, Milton Keynes recycles only 30% of its household waste. On the other hand, the promotion of biodiversity is a key part of parks and watercourse management. The parks drain runoff into a system of 13 balancing lakes, designed to hold rain water and release it slowly into the river system to reduce flood risk.

Although planned at a time when sustainability was not in the planners' vocabulary, Milton Keynes can be seen as a pioneering venture. Its achievements have been considerable, but its shortfalls are also evident.

Crossing the Atlantic takes us to perhaps the most unlikely city — Chattanooga — which developed in the nineteenth and early twentieth centuries as an important railway junction and a centre for manufacturing textiles and metals.

Case study 32   CHATTANOOGA

A role model for the USA?

In 1969, the US Environmental Protection Agency presented a special award to Chattanooga, Tennessee for being 'the dirtiest city in America'. Just over 21 years later, it was applauded as the nation's best 'turn-around story'. How did this change come about and who was responsible?

Credit for re-imaging Chattanooga has to go to a partnership involving the city authority, the Lyndhurst Foundation (an NGO sponsored by Coca-Cola) and many individuals. This partnership has tackled head-on a complex of interlinked issues, such as affordable housing, public education, transportation alternatives, urban design, conservation of natural areas, parks and green areas, air and water pollution, recycling, job retraining and riverfront redevelopment.

Part of the success has been due to using a mix of top-down and bottom-up measures. Top-down strategies included the Clean Air Act, which forced city manufacturers to invest heavily in pollution-control equipment. Bottom-up measures included the Vision 2000 programme, which called for all citizens to visualise the city as they would like it to become. From the huge number of responses emerged a remarkably progressive agenda, resulting in the following major developments:

- Some 10 km of waterfront along the Tennessee River have been made into an imaginative urban park, cutting through the heart of the city. It is made up of playgrounds, spaces for outdoor performances, fishing piers and shaded walkways. The park has played a vital part in transforming the once rundown downtown area.
- In the downtown area, the Tennessee Aquarium is a magnet, and an **anchor development**. Around it, old warehouses have been converted into smart shopping malls and other buildings have become affordable apartments and restaurants.
- To move around in the revitalised downtown area, a unique electric shuttle bus service is in operation. It acts as a park-and-ride system, capable of moving a third of the city's downtown commuters, at a tenth of the cost of a diesel vehicle system. Chattanooga now claims to have become the electric bus capital of the world. The city manufactures 22-seater electric vehicles that are marketed all over the USA.
- The Chattanooga Neighborhood Enterprise has renovated well over 3000 units of inner-city housing.

Other projects in the pipeline include the construction of a zero-emissions eco-industrial park and a grass-roofed convention centre.

In turning Chattanooga around, sustainability has been enhanced, particularly in terms of outputs (Figure 52). The environment has benefited from substantially reduced industrial and vehicle emissions, as well as from increased green space. Within the city itself, all residents have benefited from the renovation of housing and of the downtown area.

One of the most influential ventures in city sustainability is in an LEDC — Curitiba in southern Brazil. Studies of this city — the capital of the Brazilian state of Parana — feature in many geography textbooks. This city has become a beacon venture for all those keen to make cities more sustainable. Key elements include:

- encouraging public participation, particularly involving the poor
- using low-cost technology
- promoting cheap and efficient public transport
- sorting, selling and recycling waste

More than 80 cities around the world, from Bogotá to Seoul, are now following the Curitiba blueprint of a city managed by common sense.

Case study 33 takes us to the other side of the world, to Putrajaya in Malaysia. The contrast could not be stronger in the sense that, although the success of Curitiba involves getting back to basics, Putrajaya is looking to the latest technology.

PUTRAJAYA

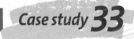
Case study 33

Malaysia's new garden capital city

Malaysia has recently joined the ranks of countries, such as Australia, Brazil and the USA, that have deliberately created a new capital city. In 1995, the authorities decided to move the national government function 25 km from the increasingly congested Kuala Lumpur (KL), to a greenfield site to the south. KL remains the country's financial and commercial centre. The new capital city, Putrajaya, should be complete by 2010, by which time its population will have reached 300 000.

Putrajaya is located within a high-tech zone, 15 km wide and 50 km long, known as the Multimedia Super Corridor (MSC) — a Malaysian 'silicon valley'. This stretches southwards from the centre of KL and includes KL International Airport, Putrajaya and Cyberjaya, a 'smart' city specialising in education, research and high-tech business. The 'spine' of the MSC is an electronic superhighway, or fibre-optic network, that provides high-speed computer links.

The undulating terrain of Putrajaya's site has been substantially modified to create a large lake surrounding an artificial island. The lake plays an important part in flood and pollution control, as well as providing aesthetic and recreational value. Just under 40% of the 500-hectare site will be left as green space.

Putrajaya is divided into two major areas: the core area and the peripheral area. The core occupies the island in the lake and is divided into the following five precincts linked by a 4 km boulevard:

- a government precinct, comprising the prime minister's office, parliamentary buildings, embassies, banks, security houses and media organisations
- a commercial precinct, where most offices will be located, a huge shopping centre and a small number of residential units
- a civic and cultural precinct — the symbolic city centre
- a mixed development precinct, involving a convention centre, sports academies and water-based recreational activities
- a sports and recreation precinct, with a theme park, low-density sports complexes, a forest park and educational institutions

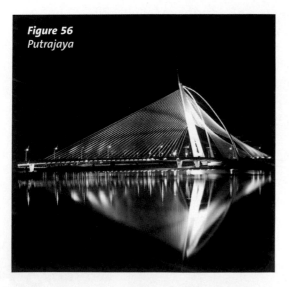

Figure 56
Putrajaya

The peripheral area comprises 15 precincts of various sizes, 12 of them residential neighbourhoods, where various houses are planned for all income groups. Most precincts will have community and neighbourhood centres, parks, places of worship, schools, hospitals, shopping centres, mosques, multipurpose halls, educational facilities and many other state-of-the-art public amenities — all linked by intelligent communication systems.

Putrajaya aims to be 'an indigenous city with a modern look'. Its land use, transportation system, infra-structure, housing, public amenities, parks and gardens have a distinctive pattern, based partly on the experiences of other new town ventures in the UK and the USA. Where Putrajaya differs is in its use of modern communications technology. It is one of a new breed of knowledge cities (*Case study 36*).

An important question concerns Putrajaya's sustainability. The running efficiency of the city should mean relatively low rates of resource consumption. The environmental impacts are considerable, as the city's development involves utilising and modifying a greenfield site. However, high levels of recycling and better flood and pollution control should be seen as environmental positives. Within the city, a comfortable and quality lifestyle is promised for all residents. Critics of the project, however, say that the small amount of low-cost housing means a degree of exclusion. As a consequence, Putrajaya will be a city enjoyed only by the more affluent.

22 **Question**

Read *Case study 33* again. Compare Putrajaya with one of the following planned capital cities: Canberra, Pretoria or Brasília.

Guidance

You could compare them in terms of reasons for building and design.

Website: **www.putrajaya.net.my**

The last city to be considered in the developing world is Shanghai, which is set to become the largest in the world by the end of the twenty-first century.

SHANGHAI

Case study **34**

Becoming eco-friendly

Shanghai, the powerhouse of China's economic miracle (*Case study 5*), has already recognised that a key aspect of sustainability is transport. The major challenge is how to reduce the energy demands and greenhouse gas emissions of city transport, while maintaining mobility within the city. Key elements in Shanghai's eco-friendly transport system are:

- a network of 11 metro or light railway lines, totalling 325 km in length
- a bus-based mass transit system integrating with the metro lines
- a railway link to the new international airport, which involves the world's first commercial magnetic levitation trains, capable of reaching 550 kmh
- reducing the numbers of cars on Shanghai's roads, mainly by raising licensing fees and restricting access to the city centre
- an electronic guidance system that helps to avoid congestion and keeps road traffic flowing

Although many Western cities encourage cycle paths as part of a more eco-friendly city transport system, the 9 million cyclists in Shanghai could face a ban from major roads, as the authorities struggle to ease congestion and appease the rising car-owning middle classes.

In addition to making its transport system more sustainable, Shanghai is planning a huge expansion in the form of a new city — Dongtan (Figure 57), which will be the size of Manhattan in New York. The city will be built at the eastern end of Chongming, a large island in the Yangtze River delta. The site is one of the last big undeveloped spaces in the Shanghai area. However, it is also an important feeding station for migrating wetland birds. The waters here are rich in aquatic resources. It is claimed that Dongtan will be the world's first genuinely eco-friendly city, powered by renewable energy resources (mainly hydroelectric power). The city will be as close to being carbon-neutral as possible. It has been designed by a British company and will be completed by 2010, in time for Shanghai's hosting of the World Expo.

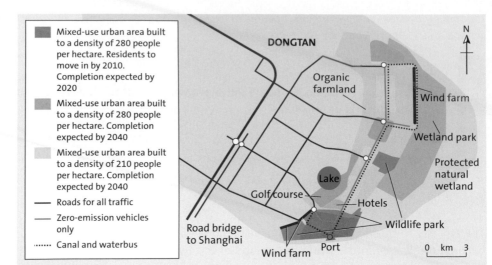

Figure 57 Dongtan: construction of an eco-city

Mixed-use urban area built to a density of 280 people per hectare. Residents to move in by 2010. Completion expected by 2020

Mixed-use urban area built to a density of 280 people per hectare. Completion expected by 2040

Mixed-use urban area built to a density of 210 people per hectare. Completion expected by 2040

—— Roads for all traffic

—— Zero-emission vehicles only

······ Canal and waterbus

DONGTAN

Organic farmland

Wind farm

Wetland park

Protected natural wetland

Lake

Golf course

Road bridge to Shanghai

Wind farm

Port

Hotels

Wildlife park

0   km   3

N

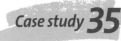

23

Using case studies

Question

Read *Case study 34* again. Do you think Dongtan will be as eco-friendly as is claimed?

Guidance

Think about the energy requirements of high-rise developments. What other details in the case study might you question?

Website: **www.dongtan.biz/english/zhdt**

London's opportunities

Our global journey in search of a sustainable city brings us back to where we started — the UK. To mention London in connection with sustainability may seem to be taking things a little too far. After all, London is a highly dynamic world city with a heavy ecological footprint — it is not a sustainable city (see Figure 53). However, London has two major opportunities in the next 20 years to change its dismal record: a major planned expansion of London in an easterly direction (often referred to as the Thames Gateway); and London's hosting of the 2012 Olympic Games. Part of the infrastructure will be built on a brownfield site at Stratford in east London — the 'portal' to the Thames Gateway.

Case study **35**   LONDON'S CHALLENGE

Moving towards sustainability

The 2004 London plan — the first for nearly 30 years — sets out the vision to make London a good example of a sustainable world city. The aim is to ensure that London's expected growth (0.8 million over the next 10 years) is accommodated in a sustainable way. This growth is only part of a wider growth scenario involving most of the southeast. The challenge is where to locate the extra homes and new jobs for this growth. Four key growth areas have been identified (Figure 58):

- the M1 corridor of Milton Keynes, Bedford, Corby and Northampton
- the M11 corridor running from London through Stansted to Cambridge
- Ashford in Kent
- the Thames Gateway

It is the last of these that is most immediately linked to London's growth.

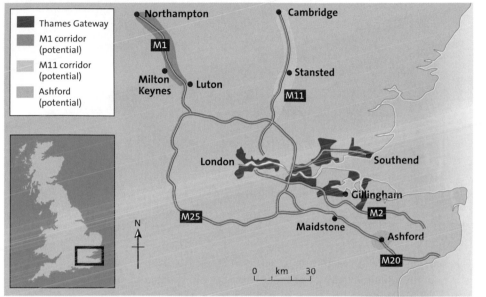

Figure 58 Four key growth areas in the southeast

The Thames Gateway project involves a band of land 30 km wide and stretching 65 km from east London (Figure 59). It represents a major eastward extension of London along the Thames Estuary, stretching from London Docklands to Southend in Essex and to Sittingbourne and Sheerness in Kent. It will be Europe's largest and most ambitious regeneration project — many of the 120 000 new houses and 180 000 jobs will be located on 3000 hectares of brownfield sites within the Gateway. This growth is planned to take place in an area that already accommodates 1.6 million people and around 500 000 jobs.

This ambitious project is based on both need and opportunity:

- The riverside corridor once contained many land-extensive industries (former docks, warehouses, factories and quarries) that used to serve London and the southeast. Their decline has left large-scale derelict sites, depressed local towns and deprived communities.
- The area is located close to the economic heart of London and on the route to the increasingly important markets in northwest Europe. It is hoped that economic success in this area will help meet the challenge of providing more social and affordable housing.
- Lifestyle changes (more young people leaving home, higher rates of divorce, more single-person households and one-parent families) are stimulating a greater demand for housing than the actual rate of population growth.

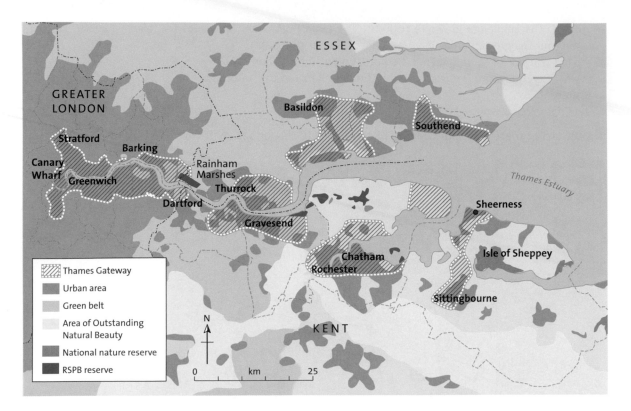

ESSEX

GREATER
LONDON

Basildon

Southend

Stratford
Barking
Canary
Wharf
Greenwich
Rainham
Marshes
Thurrock
Dartford
Gravesend

Thames Estuary

Sheerness

Isle of Sheppey

Chatham
Rochester

Sittingbourne

N

KENT

Thames Gateway
Urban area
Green belt
Area of Outstanding
Natural Beauty
National nature reserve
RSPB reserve

0   km   25

Figure 59 The Thames Gateway proposals

Five areas have been designated to accommodate the planned expansion:

■ a new metropolitan district of Inner East London, capable of sustaining 48 000 new houses and focused on Stratford and the Lower Lea valley. This will also be the site of the 2012 London Olympics. The Olympic Park, comprising a new stadium and athletes' village, will occupy a 200-hectare site (Figure 60). After the Games, it is proposed that half of the 3600 apartments and townhouses will be used as social housing and the rest sold privately.

Figure 60 An artist's impression of the Olympic Park in London

■ an area known as Outer Riverside (27 750 homes), lying to the south of the River Thames between the Greenwich peninsula and Woolwich

■ Mid Gateway City and Ebbsfleet (47 300 homes) — a key growth area with a station on the new Channel Tunnel link from Folkestone to London St Pancras

support@london2012.com

Contemporary Case Studies

- Medway City (40 610 homes)
- South Essex towns (15 900) — Thurrock and expansions to Basildon and Southend.

This is a unique opportunity to turn the area's heritage of unrestricted urban growth — with its breakers' yards, derelict land and deprived communities — into a flagship sustainable development. It will need a partnership of public and private enterprise. Major private sector parties in the project range from real estate developers to the Royal Society for the Protection of Birds (RSPB).

Specific moves in the sustainability agenda include:

- providing affordable housing for key workers
- promoting housing densities of 30–50 dwellings per hectare, to minimise the use of greenfield sites
- reducing the risk of flooding — much of the Thames Gateway lies within the natural tidal floodplain of the River Thames
- 'greening' the Gateway by protecting and enhancing important wildlife areas and providing public open spaces (Figure 59) — Rainham Marshes in Essex will be a flagship ecological and leisure scheme led by the RSPB
- providing a range of leisure facilities, such as sports venues, exhibition sites and conference centres

One of the biggest challenges will be dealing with the flood risk. Much of the planned growth will be located on the Thames floodplain. Homes built below the Thames Barrier will be at risk, as well as several thousands of new homes above the Barrier in Barking, Dagenham, Havering and Woolwich. The developers are already talking about extra measures to lessen the risk, including using the ground floors of houses for car parking rather than living space, raising ground levels and building secondary embankments.

Insurance cover for new homes in the Thames Gateway will depend on adequate flood defences being put in place. Ironically, the Association of British Insurers is campaigning to stop planners giving permission for homes on areas where the risk of flooding is high.

The 2012 Olympic Games will turn the world's spotlight on the Thames Gateway project. The challenge that faces London is how to become a more sustainable global city.

Websites: **www.thames-gateway.org.uk**
www.olympic.org/uk/news/olympic
www.lda.gov.uk
http://olympics.newham.gov.uk

24 *Using case studies*

Question

Moving towards sustainability requires a number of different actions.
(a) Read *Case studies 30–35* again and make a list of these actions.
(b) Which do you think are the most important?

Guidance

The bullet points on p. 84 may help you to answer the question.

International efforts

Table 5 *Recent global initiatives towards a better urban future*

The cites examined in this book demonstrate the roles of national and local policies. However, Table 5 shows how international efforts are increasingly being made to improve physical and socio-economic environments.

Date	Initiative
1972	■ UN Conference on Human Development held in Stockholm ■ The first meeting of the international community to discuss global environment and development needs ■ UN Environmental Programme (UNEP) established
1980s	■ UN set up World Commission on Environment and Development (Brundtland Commission). In 1989, it produced a report called *Our Common Future*. Sustainability defined as 'meeting the needs of present generations without compromising the ability of future generations to meet their own needs'. This provided the framework for Agenda 21 and for the principles set out in the Rio Declaration on the Environment and Development
1992	■ Earth Summit at Rio de Janeiro, where 150 nations signed a 'blueprint for sustainable development' called Agenda 21 and the Rio Declaration. They also launched the Local Agenda 21 ('global problems, local action') process
1992 onwards	■ Numerous initiatives by cities globally to fulfil their commitment to Local Agenda 21 and sustainability. These span high-profile pollution monitoring and traffic management in London, CCTV and crime-busting initiatives in many MEDC cities, and lower-profile waste disposal schemes by the Zabbaleen in Cairo. Some cities, such as Chattanooga in USA and Curitiba in Brazil, promote themselves as models of sustainability
1993	■ European Sustainable Cities project, financed by the European Union. Local authorities given guidance on how to plan long term in more sustainable ways
1994	■ Global Forum at Manchester, aimed at developing action plans for sustainable city growth. Fifty cities were represented, but the forum was criticised for superficial 'greening' policies.
1996	■ UN Conference Habitat II held in Istanbul. Over 150 countries discussed how to curb urban growth, how to prioritise housing and sanitation, and how to promote public–private partnerships. The UN is proactive in promoting more effective environmental planning and management via its UN Habitat and Sustainable Cities Programme. *The Gaia Atlas of Cities* was published, which highlighted the differences between linear and circular metabolisms in city systems
2000	■ Eight Millennium Development Goals were adopted by UN member states. The goals range from poverty reduction, health and gender equality to education and environmental sustainability. They are scheduled to be achieved by 2015
2002	■ Earth Summit at Johannesburg reaffirmed the significance of several key indicators for sustainability
2002	■ Greater London Authority published its *City Limits Report*. This assessed the sustainability of the city's consumption of resources and the size of its ecological footprint. It highlighted the role to be played by technology in producing a more sustainable city in the future
2005	■ UN Sustainable Cities programme met to assess progress on the Millennium Development Goals. Slum upgrading and prevention are critical to achieve the goals and targets in a world where half the global population lives in urban areas, and a third of them — about 1 billion people — live in slums

Problems

The concept of sustainable cities is spreading slowly around the world. Although this is encouraging, a truly sustainable city may prove to be nothing more than a pipedream. Despite the most stringent measures, all cities have inevitable internal problems. Figure 61 shows that there are four internal goals to be met:

- a sound economic base
- an environmental quality acceptable to city residents
- an even allocation of resources and opportunities across city society
- involvement of the community in decision making about the city's future

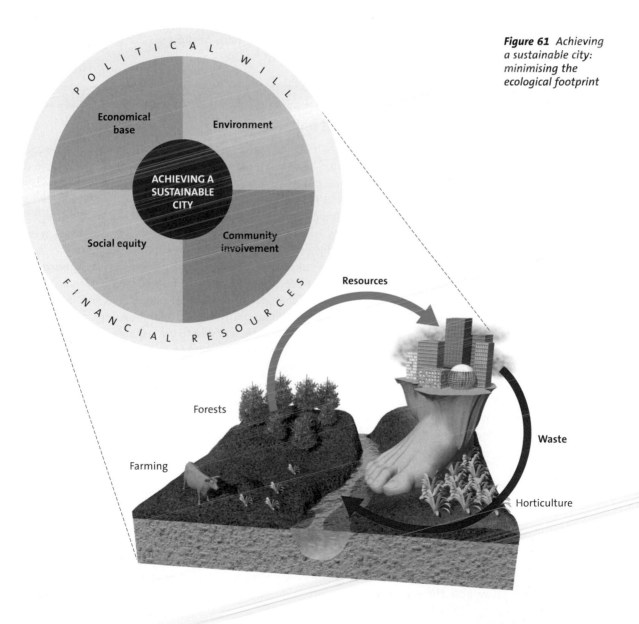

Figure 61 *Achieving a sustainable city: minimising the ecological footprint*

The path to these goals lies in:

- having the political will to make what may be painful changes in the short term, such as congestion charging
- allocating adequate financial resources
- reducing the external ecological footprint

There are huge financial implications related to sustainability, so where exactly a country or city is located along the development pathway may have a direct bearing on sustainability. Comparisons of Chattanooga and London with Addis Ababa and Dhaka support this view.

The role model for the global community is Curitiba, a city that started to have an integrated policy for sustainability at a time when Brazil was only just acquiring its NIC status. Curitiba illustrates the importance of political will and community involvement, rather than simply economic status and financial resources. *Case study 36* poses the inevitable question — does the city have a future?

Case study 36 — DOES THE CITY HAVE A FUTURE?

Looking into the crystal ball

Sir Peter Hall, an eminent urban planner, predicts that cities, particularly what he calls 'creative cities', do have a future. Although information technology has been unable to prevent the decentralisation that has taken place over the last 30 years, there will always be a hard core of activities that require face-to-face contact. These will remain rooted in the city, but not necessarily in their present locations within the built-up area. Hall argues that, in spite of their rigid physical form, many cities have shown how capable they are at adapting to new economic forces. From being centres of manufacturing, they are now centres for advanced services — unimaginable even 20 years ago. Industrial buildings used in manufacturing goods have been recycled and transformed into high-tech offices. Sites for dirty industries have been changed into industrial estates for clean industries, services and new shopping centres. Where once fine Victorian buildings stood in city centres, now there are glass-fronted structures adapted to the technical demands of the financial services industry. Cities are dynamic, changing to new needs as they emerge. Hall is confident that they will come up with creative solutions to today's problems, such as traffic congestion and the rich–poor divide, just as they have done in the past.

There are many people who, although they agree with Hall, define city futures in more graphic terms as being a basic choice between sprawl, decentralisation and decline on the one hand and high density, high value added, healthy cores and success on the other.

Knowledge cities

Many of today's creative cities are those that are involved in the emerging knowledge-based economy. Knowledge cities have an economy driven by high value-added exports, created through research, technology and brainpower. In other words, they are cities in which both private and public sectors value and nurture knowledge, and spend money on supporting knowledge dissemination and discovery (i.e. learning and innovation). They harness this knowledge to create new products and services that add value and create wealth. There are currently 65 urban development programmes worldwide, designated as knowledge cities. Putrajaya (*Case study 33*) is one of them, and Zhuhai in south China is another popular example.

Participation in the knowledge-based economy is reflected in the attractively designed research and development campuses and science parks in the built-up area. Areas of high-quality residential development are located nearby. These are just part of the overall package of vibrant 'live, work and play' environments needed to attract and retain the talented, educated people who are the key workers in the knowledge-based economy.

These knowledge cities are the next generation of 'silicon valley' urban complexes. Only those countries and cities keen and able to embrace economic globalisation will participate in this new phase of the post-industrial era. A key question is: how sustainable will such cities be?

Looking to the future, there is now one inescapable fact — global warming. What sort of future does global warming promise for today's cities and for sustainability? To answer this question, three important observations should be made:

- Many cities located close to sea-level will have to confront an increased risk of flooding. In some cases, the threat may be so great that low-lying districts will have to be abandoned. It is estimated that at least 14 of today's mega-cities fall into this category. City authorities and businesses are gradually waking up to this new scenario. Planning is increasingly about 'mitigation' strategies that hope to reduce the severity of the threat.
- Global warming will impact on the city metabolism by increasing inputs of water and energy (particularly for air conditioning) and as a consequence outputs will be raised. The urban 'heat island' is likely to become larger and more conspicuous.
- Worsening conditions in rural areas of marginal agriculture will generate a growing exodus of environmental refugees. Will cities be able to cope with this huge influx of rural migrants?

Finally, it should not be forgotten that cities are major contributors to global warming. This fact alone requires that the top item on any city sustainability agenda should be the reduction of carbon emissions. Failure to reduce emissions will make the future of cities and urbanisation less certain.

Websites: www.earth-policy.org/Updates/Update33.htm
www.ehs.unu.edu/print.php?page=12_October_-_UN_Disaster_Day
www.ecopolis.com.au/
www.timesonline.co.uk/article/0,,2-2100776,00.html
www.sfgate.com/cgi-bin/article.cgi?f=/n/a/2005/05/29/state/n122800D35.DTL

Examination advice

Most A-level geography examination questions ask candidates to weave together theoretical ideas (such as global cities, suburbanisation or sustainability) and case study material.

The case study material may be short, supporting examples, but often a whole answer may be built around a single case study. This book contains information about a number of countries and their leading cities at different points along the urbanisation pathway. You should be able to use these case studies for a variety of in-depth questions. Also included are shorter descriptions of other urban phenomena, which you might be able to use as passing examples to show breadth of knowledge.

Table 6 Question commands and the examiner's case study expectations

| Task or command* | Required case study detail | | | |
	Name	Support	Compare	Examine (a particular situation or statement)
Typical question	Give an example of a post-industrial city	With the use of examples, show how cities modify the natural environment	Compare LEDC and MEDC cities in terms of the location of poor housing	With reference to a named country, examine the main factors affecting the rate of urbanisation
Case study expectation	Do no more than name an appropriate city, e.g. Bradford, Manchester, Chattanooga	Do no more than name-drop. Refer to three different examples, e.g. pollution of an industrial city (Detroit), coastal reclamation (Southampton) and changes to hydrology (Milton Keynes)	Use two contrasting cities, e.g. Dhaka and London; compare the inner- and fringe-city locations typical of LEDC cities with inner-city slums and edge-of-city 'sink' estates of MEDC cities	Give detailed information about one country; use Figure 14 as a checklist of the factors you could consider

*A word of caution — watch out for those questions that do not specifically ask for examples, but nonetheless expect them. For example:

- Suggest reasons why traffic is a major urban issue.
- 'Shanty towns are a sign that all is not well with the LEDC city.' Discuss.
- Examine the costs and benefits of intensifying suburban development.

If you are in doubt as to whether examples are required in your answer to a question, it is better to give some rather than none.

Candidates often penalise themselves by not reading the question carefully and by not responding to its precise demands. Examination success is based on two abilities:

- Recognising the precise context of the question. The challenge here is to find an example or case study that is appropriate to the question topic. Is this a question just about LEDCs or MEDCs, or is it a more open, global question?
- Recognising the command word used in the question and responding accordingly. Table 6 demonstrates that there are at least four different question scenarios: name, support, compare and examine. They are ranked in terms of what degree of case study detail is expected, which ranges from simple name-dropping to detailed knowledge and understanding of a particular situation.

In the remaining sections, we look at five different examination contexts that require the use of case studies. They include the four shown in Table 6 and apply mainly to unseen examination papers. The fifth is where the geography specification requires you to undertake an enquiry into a set topic and to submit a report by a given deadline (e.g. Paper 6475/02 of the Edexcel Geography B specification). Therefore, the contexts range from simply naming an example of a particular situation, through using case study material for support purposes, to an extended and detailed use of case studies.

Naming examples

This requires nothing more, than being able to cite one of your case studies as a relevant example. The test here is basically one of appropriateness. In part (b) of the sample question below, you need not confine yourself to symptoms shown by the country you named in (a).

---

25 | **Using case studies**

Question
(a) Name one country that is experiencing counterurbanisation.
(b) Identify three symptoms of counterurbanisation.

Guidance

(a) The UK; most Western European countries
(b) ■ The largest city loses population
  ■ Low-order and middle-order cities begin to take off
  ■ New life in rural areas

See Counterurbanisation (p. 23).

Naming and using a supporting example

The next level up requires that you do more than just name an appropriate example. You are also expected to demonstrate knowledge and understanding by providing some detail.

26

Question

With reference to a named city, explain what is meant by urban re-imaging.

Guidance

Re-imaging involves the efforts of a city to shake off a negative and dated image for a more positive one that is likely to make the city more attractive to people and business. Look back at *Case study 19* to see what Bradford is doing and use this information to give your answer a ring of authenticity.

Using case studies comparatively

Essay-type questions that require the comparative use of case studies are popular with chief examiners. Much of the challenge of such questions hinges on selecting appropriate case studies. In some instances, the choice is fairly obvious and restricted. In others, there may be more choice and options than you might first think from initially reading the question.

The trouble with all examination questions that ask you to compare situations (as represented by appropriate case studies), is that some candidates believe that all they have to do is to rehash each in turn. This leaves examiners to draw their own conclusions as to whether or not the situations are similar. In short, the question is not answered and relatively few marks can be gained. In planning effective answers to most comparative types of question, it is necessary to interleave references to your chosen case studies.

Question

With reference to two contrasting examples, evaluate the role of government policies in influencing the distribution of urbanisation.

Guidance

From the start, you should grasp that this is a two-sided question, in that influence can be both encouraging and discouraging ('stick and carrot'). It may be that the obvious thing to do here is to compare a country that has tried to suppress and redirect urbanisation with one that has tried to encourage it — the UK (*Case study 11*) and China (*Case study 5*) would be good choices.

Take a closer look at the question. There is nothing to say that you must use two different countries in your answer. You could quite legitimately answer this question using only the UK as your case study country. What you would need to do is to contrast those policies designed to curb the further growth of London (e.g. those relating to the green belt) with those that have tried to remove some of the capital's overspill elsewhere (e.g. new and expanded towns). To give your answer a topical feel, you may decide to finish by taking a look at the Thames Gateway proposals. Even here, it is possible to pursue the 'stick and carrot' idea, in that there is a mix of strategies, both protecting important wildlife areas, as well as encouraging the reuse of brownfield sites.

Building an essay around a single case study

A successful attempt to answer any question that starts 'With reference to a named country...' will require:

- the choice of an appropriate country — does it illustrate the particular topic? Is it in the right development context?
- sufficiently detailed knowledge of that country in relation to the particular question
- you to resist the temptation to set down all that you know about your chosen country (the 'everything but the kitchen sink' approach), but instead harness only those aspects that are directly relevant

28

Question

For a named country with a primate city, examine the factors that may have encouraged that primacy.

Guidance

- Name your choice of country and its primate city. Give some indication of that primacy (e.g. how many times larger it is than the second city) and whether it is currently increasing or decreasing.
- Organise your discussion of possible factors under headings. For example:

Physical factors include:
- small extent; primate city readily accessible from all parts of the country

Historic factors include:
- city commanded some initial or comparative advantage
- strong central government

Demographic factors include:
- high rate natural increase in population
- continuing large volume of net in-migration from other parts of the country and overseas

Economic factors include:
- successful economy
- strong attraction to investment
- superior high-order services

Social factors include:
- appeal as 'the centre of the universe'
- superior quality of life
- better opportunities

Having run through the range of possible factors, you need to include a final paragraph that attempts some sort of evaluation. What appear to have been the most significant factors in the case of your chosen country?

Planning an extended essay involving a range of case studies

The standard advice on essay planning should be followed, namely that the essay should have a three-part structure — a brief **introduction** and an equally brief **conclusion** (one paragraph each), separated by a series of paragraphs (the **expansion**) that develop your argument or discussion points. It is here that you should incorporate supporting examples and case studies.

Question

Do you think that cities can ever become sustainable? Justify your viewpoint with reference to examples.

Guidance

Introduction

Start by establishing what sustainability means in relation to cities. What are the key issues that need to be tackled in order for a city to become more sustainable?

Expansion

Select sustainable city projects at different points along the urbanisation pathway — LEDC (e.g. Curitiba or Shanghai), NIC (e.g. Putrajaya) and MEDC (e.g. Chattanooga). This allows you to raise the question of whether city sustainability is more likely to be achieved in MEDCs than in LEDCs. MEDCs may have the resources and technology to make some impact, but their starting point is a much heavier ecological footprint. Therefore, it may be that there is more promise in LEDCs with appropriate use of intermediate technology.

Conclusion

You might conclude that cities can be made more sustainable than they are today, but can any city ever really be wholly sustainable? All cities, no matter how eco-friendly, will always have an ecological footprint.

Creating diagrams and sketch maps to illustrate case studies

An essential skill of any geographer is being able to select information from the data available. A similar skill is to create simplified sketch maps, to highlight or summarise key aspects of a geographical situation. The sketch maps in Figure 62 illustrate how these skills can be demonstrated with respect to one of the most popular student topics: urban features, processes and change. Figure 62 is the base map. From this source, a student has drawn annotated sketch maps of the main land use characteristics of Shanghai (Figure 63) and differences in the quality of life within the city (Figure 64).

30

Question

Draw a sketch map similar to Figures 63 and 64 (page 106) to show the processes that have moulded the structure of Shanghai as shown in Figure 62. Use and locate the terms urbanisation, suburbanisation, gentrification, regeneration, brownfield development and greenfield development.

Guidance

Use colours, provide a key and add annotations rather than just labels.

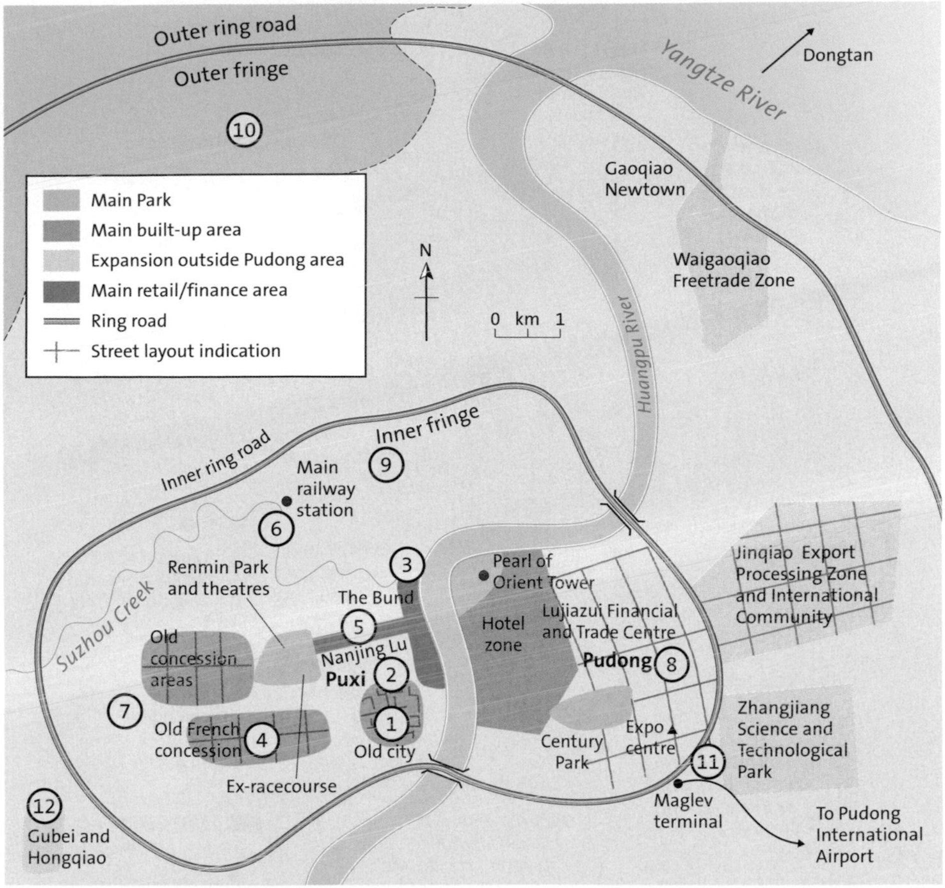

Figure 62
Basic structure of Shanghai

Outer ring road
Outer fringe

Yangtze River → Dongtan

⑩

Gaoqiao
Newtown

Waigaoqiao
Freetrade Zone

	Main Park
	Main built-up area
	Expansion outside Pudong area
	Main retail/finance area
—	Ring road
+	Street layout indication

N

0 km 1

Huangpu River

Inner ring road
Inner fringe

⑨

Main
railway
station

Suzhou Creek

⑥

③

Pearl of
Orient Tower

Jinqiao Export
Processing Zone
and International
Community

Renmin Park
and theatres

The Bund

⑤

Hotel
zone

Lujiazui Financial
and Trade Centre

Old
concession
areas

Nanjing Lu

Puxi ②

Pudong ⑧

Zhangjiang
Science and
Technological
Park

⑦

①

Old city

Century
Park

Expo
centre

⑪

Old French
concession

④

Ex-racecourse

Maglev
terminal

To Pudong
International
Airport

⑫
Gubei and
Hongqiao

① **Nanshi —
old city** — 4 km²; ancient artisan core based on medieval
walled city; tiny alleyways and high-density
nineteenth-century housing and street
markets, festooned with washing; lacks
sanitation; increasingly a hot spot for tourists

② **Puxi —
central city** — Mixture of residential and retailing; high
degree of foreign influence in wide streets and
European architecture; known as the 'Paris of
the East' in the nineteenth century

③ **The Bund** — 1.5 km belt of large, prestigious buildings
along the Huangpu River opposite Pudong;
developed by foreign investors of nineteenth
century; the original CBD with banking,
financial and retailing zone; a tourist hot spot;
much renewal and some conservation of older
buildings plus more controversial redevelopment into lower-density residential units; new
flood defences on site of original harbour have
been incorporated into a promenade; bund
literally means embankment

④ **Old French
concession** — Expensive tree-lined residential streets with
some 1930s mansion blocks (first in Asia) now
subdivided into smaller units; embassy quarter
with fashionable restaurants and boutiques;
a major night life focus and tourist hot spot

⑤ **Nanjing Lu** — The Golden Mile of retailing, mainly Chinese
outlets, the busiest shopping district in China
with 1.7 million shoppers each day; similar to
downtown Hong Kong, it is partly pedestrian-
ised and includes China's largest store

⑥ **Zhabei — railway
station quarter** — Once the American quarter, by the 1930s it had
become one of the poorest sectors of the city
called 'Little Tokyo'; now a zone of poor migrant
workers housed in cramped slums; also a red
light district

⑦ **Ulumuqi Road** — Manhattan-style redevelopment scheme to lower
population density from 37 000 to 20 000 people
per km² by forced relocation to outer suburbs

⑧ **Pudong special
economic zone** — Until 1980s a mix of farmland, poor housing,
unemployed migrants and prostitutes; now a flagship redevelopment 1.5 times bigger than central
Shanghai (350km²); foreign investors have helped
to create a 'zoned' environment of expensive
housing, businesses and leisure facilities; 2 million
inhabitants by 2002 in the Jinqiao zone; 800 high
rise buildings in Lujiazui financial zone

⑨ **Near inner
ring road** — First major fringe area of post-1945 period; mix
of regemented housing blocks and polluting
industry; 68% of Shanghai population lives within
the inner ring road; factories are increasingly
relocated to outer ring road

⑩ **Outer fringe** — City boundary includes rural areas; a juxta-
position of mixed farms, industry and suburban
development

⑪ **Universal Studios
and Disneyland** — Proposed in Pudong to become a tourist hot spot
by 2010

⑫ **Gubei** — New planned community in 'Legoland' type
estates serving the international trade area of
Hongqiao; a major hot spot for TNCs

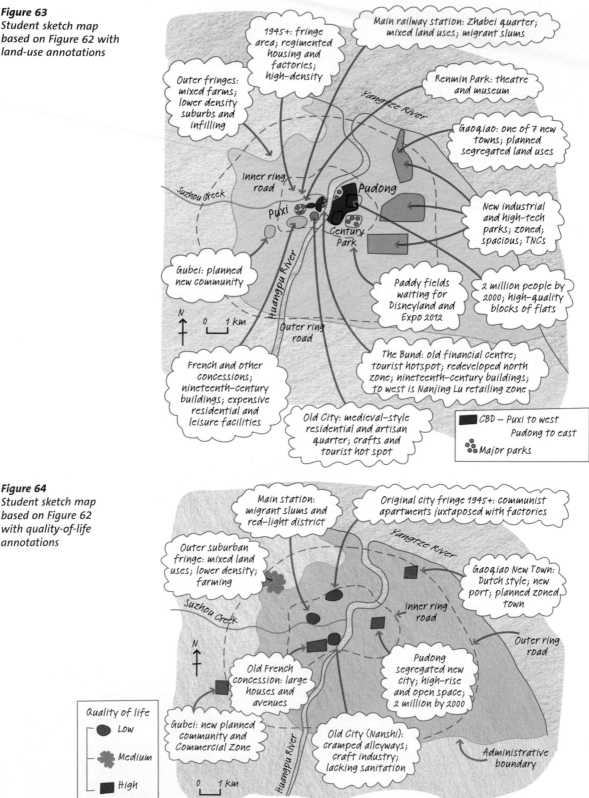

Figure 63
Student sketch map based on Figure 62 with land-use annotations

1945+: fringe area; regimented housing and factories; high-density

Main railway station: Zhabei quarter; mixed land uses; migrant slums

Renmin Park: theatre and museum

Outer fringes: mixed farms; lower density suburbs and infilling

Gaoqiao: one of 7 new towns; planned segregated land uses

Yangtze River

Inner ring road

New industrial and high-tech parks; zoned; spacious; TNCs

Suzhou Creek

Pudong

Puxi

Century Park

Gubei: planned new community

Huangpu River

Paddy fields waiting for Disneyland and Expo 2012

2 million people by 2000; high-quality blocks of flats

N   0   1 km

Outer ring road

French and other concessions; nineteenth-century buildings; expensive residential and leisure facilities

The Bund: old financial centre; tourist hotspot; redeveloped north zone; nineteenth-century buildings; to west is Nanjing Lu retailing zone

Old City: medieval-style residential and artisan quarter; crafts and tourist hot spot

CBD – Puxi to west
Pudong to east
Major parks

Figure 64
Student sketch map based on Figure 62 with quality-of-life annotations

Main station: migrant slums and red-light district

Original city fringe 1945+: communist apartments juxtaposed with factories

Outer suburban fringe: mixed land uses; lower density; farming

Yangtze River

Gaoqiao New Town: Dutch style; new port; planned zoned town

Suzhou Creek

Inner ring road

Outer ring road

N

Old French concession: large houses and avenues

Pudong segregated new city; high-rise and open space; 2 million by 2000

Quality of life
Low
Medium
High

Gubei: new planned community and Commercial Zone

Old City (Nanshi): cramped alleyways; craft industry; lacking sanitation

Administrative boundary

Huangpu River

0   1 km

Levels marking

Extended essays, essay questions and the parts of structured questions that require extended writing are marked using levels mark schemes. These schemes indicate the general characteristics of answers at different levels of achievement. The number of levels normally ranges from three to five.

Using case studies

Question

With reference to contrasting LEDCs, analyse the factors encouraging urbanisation.

Guidance

A possible levels mark scheme for this question would be:

Level 3 9–10 marks	A detailed, structured account with a clear focus on a range of factors causing urbanisation. Account is supported by detailed examples from differing types of LEDCs
Level 2 5–8 marks	Some structure to the account which provides a satisfactory insight into a limited number of factors. There is limited exemplification.
Level 1 1–4 marks	Basic ideas only. Tends to be generalised with very limited reference to the examples.

Read the student answer that follows, which was assessed as being at the lower end of the level 2 band. The following notes indicate how this answer might have been raised to a clear level 3.

Although there are similar factors influencing the type and rates of urbanisation in less developed countries such as Mexico, Malaysia, China, Bangladesh and Ethiopia, there are several individual factors too. (1)

Cities act as magnets (2) to people from surrounding rural areas. This is true in all the case studies chosen. Other factors include the decision of governments to actively encourage inward investment from abroad, such as Ethiopia in the late nineteenth century or China after the Second World War. In Addis Ababa the choice of the city for the HQ of the UN Economic Commission on Africa and the OAU is also an important factor in its growth. (3) Governments can also affect migration into cities by their policies, as shown in China. Here it was limited in the 1960s but since then the policy has changed to encourage urban development. Indeed the Pearl River Delta and lower Yangtze River (including Dongtan and Shanghai) megalopolises are growing so rapidly today because of government policy. (4)

Most rapidly growing LEDC cities have had some initial advantage, such as a coastal or river location (e.g. Shanghai or Dhaka), or industrial growth based on local resources (e.g. Baghdad and oil production). (5)

Specific factors encouraging urbanisation have been for example in Ethiopia to establish the royal family's residence in Addis Ababa after independence, which encouraged investment there. The in-migration of ex soldiers and redundant civil servants also swelled numbers. In Malaysia the decision by the government to create a new capital city, Putrajaya, has encouraged urbanisation in a previously rural area. (6)

Improvements

(1) A definition of an LEDC would be a good starting point, and a mention that there is a spectrum from the LDC of Ethiopia, LEDC of Bangladesh to the NIC of Mexico and Malaysia or RIC of China. The student has not said how these countries contrast. A statement of the main ways in which cities grow — by in-migration and natural increase — would be useful.

(2) Use of appropriate geographical terminology —e.g. the term 'centripetal' — would lift the answer to a higher level.

(3) What is the rate of growth? Specific facts and figures are needed to back up general statements.

(4) A mention of the differences between a communist state and a more democratic country with greater freedom of movement would help to explain how the Chinese and Ethiopian governments have been such a dominating factor.

(5) Baghdad has been introduced here but not in the mini introduction.

(6) Direct and indirect factors could be differentiated — direct meaning specific government policies; indirect would be the growth of industries and TNC investment acting as a magnet to inmigrants. The discussion needs to be more focused on the process of urbanisation than individual cities. Some sort of conclusion would raise the overall value of this essay.

Very little is guaranteed in this world, but grasping and applying the advice contained in this chapter should have a positive outcome. It is sometimes easy to forget that the subject of geography is about the real world. The more of the real world that you can incorporate in your examination work, the more you are likely to convince and impress the examiner that you have a sound knowledge and understanding of today's world. In short, a good dose of reality in the form of relevant case studies and examples can work wonders when it comes to raising AS and A2 geography grades.

Index